SIDNEY M. LEVY

CONSTRUCTION DAILY LOG

Second Edition

McGraw-Hill
New York Chicago San Francisco Lisbon London
Madrid Mexico City Milan New Delhi San Juan
Seoul Singapore Sydney Toronto

McGraw-Hill

A Division of The **McGraw-Hill** *Companies*

1 2 3 4 5 6 7 8 9 AGM/AGM 0 9 8 7 6 5 4 3 2

ISBN 0-07-140814-2

Printed and bound by Quebecor/Martinsburg.

This book is printed on recycled, acid-free paper.

HOW TO USE THIS DAILY DIARY

The Daily Diary is an important and key element in the complete documentation of a construction project. It is often referred to when disputes, claims, or lawsuits arise either during construction or after the project has been completed. All entries in this diary must be clear, concise, and accurate and include vital information on the daily operation of the project.

On a daily basis, the following information should be recorded:

- Month, day, date—*every day*, whether work is performed or not, i.e., weekends and holidays. When there is no work, indicate "No Work" on that date
- Weather and temperature (morning, noon, and afternoon)
- List subcontractors on site that day, how many workers were in their crews, work performed, and location of work (i.e., SW corner 1st floor, Col. 5-6)
- List number of your crew on site and the operations they were performing
- Indicate what visitors came to the site including any of your company's office staff, project manager, owner, etc.
- List any inspections performed by local officials, testing labs, engineers
- Briefly describe the type of work performed that day and, if possible, the location in the building or site where work took place
- Record deliveries of equipment/materials—refer to receiving ticket numbers if possible
- Record any unusual events, occurrences, work stoppages (any reasons for stoppage, i.e., weather, labor disputes)

Remember, you may be called upon to relate the day's activities many months or possibly years after these events have taken place. After reviewing your daily diary, you may be able to refresh your memory and greatly help your company, should the need arise.

This Daily Diary also includes some handy information, such as:

1. Unified Soil Classification System to identify soil sizes and types
2. U.S./Metric Sieve sizes
3. One Cubic Yard of Concrete Will Place
4. Estimating concrete masonry units and determining brick coursing
5. Truss configurations and interpreting truss shop drawings
6. Framing and dimesional lumber sizes. Interpreting grade stamps
7. Proper handing of doors and standard frame hardware locations
8. Structural steel shop drawing weld symbols - illustrated and explained
9. Steel rebar and welded wire mesh sizes and weights
10. Metric conversion charts and useful tables

THE DAILY LOG

CONTRACTOR: ______________________________

PROJECT NAME: ______________________________

PROJECT NUMBER: ______________________________

PROJECT ADDRESS: ______________________________

____

PROJECT SUPERINTENDENT: ______________________________

PROJECT START DATE: ______________________________

CONTRACT COMPLETION DATE: ______________________________

SUPERINTENDENT'S DAILY DIARY TELEPHONE DIRECTORY

Project: ______________________________

EMERGENCY NUMBERS

Fire: ______________________________

Police: ______________________________

Hospital: ______________________________

Emergency: ______________________________

Directions to nearest hospital:

____

____

____

SUBCONTRACTOR/VENDOR TELEPHONE & FAX NUMBERS

Name: ______________________________
Contact: ______________________________
Phone: ______________________________
Fax: ______________________________

Name: ______________________________
Contact: ______________________________
Phone: ______________________________
Fax: ______________________________

Name: ______________________________
Contact: ______________________________
Phone: ______________________________
Fax: ______________________________

Name: ______________________________
Contact: ______________________________
Phone: ______________________________
Fax: ______________________________

Name: ______________________________
Contact: ______________________________
Phone: ______________________________
Fax: ______________________________

Name: ______________________________
Contact: ______________________________
Phone: ______________________________
Fax: ______________________________

Name: ______________________________
Contact: ______________________________
Phone: ______________________________
Fax: ______________________________

Name: ______________________________
Contact: ______________________________
Phone: ______________________________
Fax: ______________________________

Name: ______________________________
Contact: ______________________________
Phone: ______________________________
Fax: ______________________________

Name: ______________________________
Contact: ______________________________
Phone: ______________________________
Fax: ______________________________

Name: ______________________________
Contact: ______________________________
Phone: ______________________________
Fax: ______________________________

Name: ______________________________
Contact: ______________________________
Phone: ______________________________
Fax: ______________________________

Name: ______________________________
Contact: ______________________________
Phone: ______________________________
Fax: ______________________________

Name: ______________________________
Contact: ______________________________
Phone: ______________________________
Fax: ______________________________

Industry Associations and Trade Groups

Many construction-related organizations and trade groups serve as focal points for their related interests, and construction specifications are often sprinkled with reference to organizations that set standards for a particular trade or product.

These organizations can be important sources of technical product data, industry standards, and installation and maintenance instructions, and they are willing to assist in disseminating their information, frequently at no cost.

Here are many of these organizations, arranged by organization abbreviation, name, mailing and e-mail address, and telephone number. Since both e-mail and office addresses may change from time to time, the accuracy of all listings cannot be guaranteed.

AA — Aluminum Association
900 19th St., NW, Suite 300
Washington, DC 20006
www.aluminum.org
(202) 862-5100

AABC — Associated Air Balance Council
1518 K St., NW, Suite 503
Washington, DC 20005
www.aabchq.com
(202) 737-0202

AAMA — American Architectural Manufacturers Association
1827 Walden Office Sq., Suite 104
Schaumburg, IL 60173-4268
www.aamanet.org
(847) 303-5664

AAN — American Association of Nurserymen
(See ANLA)

AASHTO — American Association of State Highway and Transportation Officials
444 North Capitol St., NW, Suite 249
Washington, DC 20001
www.aashto.org
(202) 624-5800

AATCC — American Association of Textile Chemists and Colorists
P.O. Box 12215
One Davis Dr.
Research Triangle Park, NC 27709-2215
www.aatcc.org
(919) 549-8141

ABMA — American Bearing Manufacturers Association
(Formerly: Anti-Friction Bearing Manufacturers Association)
1200 19th St., NW, Suite 300
Washington, DC 20036-2401
www.abma-dc.org
(202) 429-5155

ABMA — American Boiler Manufacturers Association
950 North Glebe Rd., Suite 160
Arlington, VA 22203-1824
www.abma.com
(703) 522-7350

ACI — American Concrete Institute
P.O. Box 9094
Farmington Hills, MI 48333-9094
www.aci-int.org
(248) 848-3700

ACIL — ACIL: The Association of Independent Scientific, Engineering, and Testing Firms
1629 K St., NW, Suite 400
Washington, DC 20006
www.acil.org
(202) 887-5872

ACPA — American Concrete Pipe Association
222 West Las Colinas Blvd., Suite 641
Irving, TX 75039-5423
www.concrete-pipe.org
(972) 506-7216

ADC — Air Diffusion Council
104 South Michigan Ave., Suite 1500
Chicago, IL 60603
(312) 201-0101

AEIC Association of Edison Illuminating Companies
600 N. 18th St.
P.O. Box 2641
Birmingham, AL 35291-0992
www.aeic.org
(205) 250-2530

AFBMA Anti-Friction Bearing Manufacturers Association
(See ABMA)

AFPA American Forest and Paper Association
(Formerly: National Forest Products Association)
1111 19th St., NW, Suite 800
Washington, DC 20036
www.afandpa.org
(800) 878-8878
(202) 463-2700

AGA American Gas Association
1515 Wilson Blvd.
Arlington, VA 22209
www.aga.com
(703) 841-8400

AHA American Hardboard Association
1210 W. Northwest Hwy.
Palatine, IL 60067-1897
(847) 934-8800

AHAM Association of Home Appliance Manufacturers
20 N. Wacker Dr., Suite 1231
Chicago, IL 60606
www.aham.org
(312) 984-5800

AI Asphalt Institute
P.O. Box 14052
Lexington, KY 40512-4052
www.asphaltinstitute.org
(606) 288-4960

AIA The American Institute of Architects
1735 New York Ave., NW
Washington, DC 20006-5292
www.aia.org
(202) 626-7300

AIA American Insurance Association
1130 Connecticut Ave., NW, Suite 1000
Washington, DC 20036
(202) 828-7100

AIHA American Industrial Hygiene Association
2700 Prosperity Ave., Suite 250
Fairfax, VA 22031
www.aiha.org
(702) 849-8888

AISC American Institute of Steel Construction
One East Wacker Dr., Suite 3100
Chicago, IL 60601-2001
www.aisc.web.com
(800) 644-2400
(312) 670-2400

AISI American Iron and Steel Institute
1101 17th St., NW, Suite 1300
Washington, DC 20036-4700
www.steel.org
(202) 452-7100

AITC American Institute of Timber Construction
7012 S. Revere Pkwy., Suite 140
Englewood, CO 80112
www.aitc-glulam.org
(303) 792-9559

ALA American Laminators Association
(See LMA)

ALCA Associated Landscape Contractors of America
150 Elden St., Suite 270
Herndon, VA 20170-4831
www.alca. org
(800) 395-2522
(703) 736-9666

ALI Associated Laboratories, Inc.
P.O. Box 152837
Dallas, TX 75315
www.assoc-labs.com
(214) 565-0593

ALSC American Lumber Standards Committee
P.O. Box 210
Germantown, MD 20875
(301) 972-1700

AMCA Air Movement and Control Association International, Inc.
30 W. University Dr.
Arlington Heights, IL 60004-1893
www.amca.org
(847) 394-0150

ANLA American Nursery and Landscape Association
(Formerly: American Association of Nurserymen)
1250 Eye St., NW, Suite 500
Washington, DC 20005
www.anla.org
(202) 789-2900

ANSI American National Standards Institute
11 West 42nd St., 13th Floor
New York, NY 10036-8002
www.ansi.org
(888) 267-4783
(212) 642-4900

AOAC AOAC International
481 N. Frederick Ave., Suite 500
Gaithersburg, MD 20877
www.aoac.org
(301) 924-7077

AOSA Association of Official Seed Analysts
P.O. Box 81152
Lincoln, NE 68501-1152
www.zianet.com/AOSA
(402) 476-3852

APA APA-The Engineered Wood Association
(Formerly: American Plywood Association)
P.O. Box 11700
Tacoma, WA 98411-0700
www.apawood.org
(253) 565-6600

APA Architectural Precast Association
P.O. Box 08669
Fort Myers, FL 33908-0669
www.archprecast.org
(941) 454-6989

API American Petroleum Institute
1220 L St, NW, Suite 900
Washington, DC 20005-8029
www.api.org
(202) 682-8000

ARI Air-Conditioning and Refrigeration Institute
4301 Fairfax Dr., Suite 425
Arlington, VA 22203
www.ari.org
(703) 524-8800

ARMA Asphalt Roofing Manufacturers Association
Center Park
4041 Powder Mill Rd., Suite 404
Calverton, MD 20705
www.asphaltroofing.org
(301) 348-2002

ASA Acoustical Society of America
500 Sunnyside Blvd.
Woodbury, NY 11797
asa.aip.org
(516) 576-2360

ASC Adhesive and Sealant Council
1627 K St., NW, Suite 1000
Washington, DC 20006-1707
www.ascouncil.org
(202) 451-1500

ASCA Architectural Spray Coaters Association
895 Doncaster Dr.
West Deptford, NJ 08066
(609) 848-6120

ASCE American Society of Civil Engineers
World Headquarters
1801 Alexander Graham Bell Dr.
Reston, VA 20191-4400
www.asce.org
(800) 548-2723
(703) 295-6000

ASHES American Society for Healthcare Environmental Services
Division of the American Hospital Association
One North Franklin, Suite 2700
Chicago, IL 60606
(312) 422-3860

ASHRAE American Society of Heating, Refrigerating and Air-Conditioning Engineers
1791 Tullie Circle, NE
Atlanta, GA 30329-2305
www.ashrae.org
(800) 527-4723
(404) 636-8400

ASLA American Society of Landscape Architects
636 Eye St., NW
Washington, DC 20001-3736
www.asla.org
(202) 898-2444

ASME American Society of Mechanical Engineers
345 East 47th St.
New York, NY 10017-2392
www.asme.org
(800) 843-2763
(212) 705-7722

ASPA American Sod Producers Association
(See TPI)

ASPE American Society of Plumbing Engineers
3617 Thousand Oaks Blvd., Suite 210
Westlake Village, CA 91362-3649
(805) 495-7120

ASQ American Society for Quality
611 East Wisconsin Ave.
Milwaukee, WI 53201-3005
www.asq.org
(800) 248-1946
(414) 272-8575

ASSE American Society of Sanitary Engineering
28901 Clemens Rd.
Westlake, OH 44145
www.asse-plumbing.org
(440) 835-3040

ASTM American Society for Testing and Materials
100 Barr Harbor Dr.
West Conshohocken, PA 19428-2959
www.astm.org
(610) 832-9500

ATIS Alliance for Telecommunications Industry Solutions
(Formerly: Exchange Carriers Standards Association)
1200 G St., NW, Suite 500
Washington, DC 20005
www.atis.org
(202) 628-6380

AWCI Association of the Wall and Ceiling Industries-International
803 West Broad St., Suite 600
Falls Church, VA 22046
www.awci.org
(703) 534-8300

AWCMA American Window Covering Manufacturers Association
(See WCMA)

AWI Architectural Woodwork Institute
1952 Issac Newton Sq. West
Reston, VA 20190
www.awinet.org
(800) 449-8811
(703) 733-0600

AWPA American Wood-Preservers' Association
P.O. Box 5690
Granbury, TX 76049
www.awpa.com
(817) 326-6300

AWPB American Wood Preservers' Bureau
(This organization is now defunct.)

AWS American Welding Society
550 NW LeJeune Rd
Miami, FL 33126
www.amweld.org
(800) 444-9353
(305) 443-9353

AWWA American Water Works Association
6666 W. Quincy Ave.
Denver, CO 80235
www.awwa.org
(800) 926-7337
(303) 794-7711

BAC Brick Association of the Carolinas
(Formerly: Brick Association of North Carolina)
P.O. Box 13290
Greensboro, NC 27415-3290
www.gobrick.com
(800) 622-7425
(336) 273-5566

BHMA Builders Hardware Manufacturers Association
355 Lexington Ave., 17th Floor
New York, NY 10017-6603
(212) 661-4261

BIA Brick Industry Association
11490 Commerce Park Dr.
Reston, VA 22091-1525
www.bia.org
(703) 620-0010

BIFMA The Business and Institutional Furniture Manufacturer's Association International
2680 Horizon Dr., SE, Suite A1
Grand Rapids, MI 49546-7500
www.bifma.com
(616) 285-3963

CAGI Compressed Air and Gas Institute
c/o Thomas Associates, Inc.
1300 Sumner Ave.
Cleveland, OH 44115-2851
www.taol.com/cagi
(216) 241-7333

CAUS Color Association of the United States
409 W. 44th St.
New York, NY 10036-4402
(212) 582-6884

CBMA Certified Ballast Manufacturers Association
355 Lexington Ave., 17th Floor
New York, NY 10017
www.certbal.org
(212) 661-4261

CCC Carpet Cushion Council
P.O. Box 546
Riverside, CT 06878-0546
(203) 637-1312

CDA Copper Development Association Inc.
260 Madison Ave., 16th Floor
New York, NY 10016-2401
www.copper.org
(800) 232-3282
(212) 251-7200

CFFA Chemical Fabrics & Film Association, Inc.
c/o Thomas Associates, Inc.
1300 Sumner Ave.
Cleveland, OH 44115-2851
www.taol.com/cffa
(216) 241-7333

CGA Compressed Gas Association
1725 Jefferson Davis Hwy, Suite 1004
Arlington, VA 22202-4102
www.cganet.com
(703) 412-0900

CGSB Canadian General Standards Board
Place du Portage
Phase III, 6B1
11 Laurier St.
Hull, Quebec K1A 0S5
CANADA
www.pwgsc.gc.ca/cgsb
(819) 956-3500
Mailing Address:
Canadian General Standards Board
Sales Centre
Ottawa K1A 1G6
CANADA
(819) 956-0425

CISCA Ceilings and Interior Systems Construction Association
1500 Lincoln Hwy, Suite 202
St. Charles, IL 60174
www.cisca.org
(630) 584-1919

CISPI Cast Iron Soil Pipe Institute
5959 Shallowford Rd., Suite 419
Chattanooga, TN 37421
www.cispi.org
(423) 892-0137

CLFMI Chain Link Fence Manufacturers Institute
9891 Broken Land Pkwy., Suite 300
Columbia, MD 21046
www.baileadership.com
(301) 596-2584

CPA Composite Panel Association
(Formerly: National Particleboard Association)
18928 Premiere Ct.
Gaithersburg, MD 20879-1569
www.pbmdf.com
(301) 670-0604

CPPA Corrugated Polyethylene Pipe Association
3621 Secor Rd., Suite 320
Toledo, OH 436046
www.cppa-info.org
(800) 510-2772
(419) 241-2221

CRI Carpet and Rug Institute
310 S. Holiday Ave.
Dalton, GA 30722-2048
www.carpet-rug.com
(800) 882-8846
(706) 278-3176

CRSI Concrete Reinforcing Steel Institute
933 N. Plum Grove Rd.
Schaumburg, IL 60173-4758
www.crsi.org
(847) 517-1200

CSSB Cedar Shake and Single Bureau
515 116th Ave., NE, Suite 275
Bellevue, WA 98004-5294
www.cedarbureau.org
(206) 453-1323

CTI Ceramic Tile Institute of America
12061 West Jefferson Blvd.
Culver City, CA 90230-6219
(310) 574-7800

CTI Cooling Tower Institute
P.O. Box 73383
Houston, TX 77273
www.cti.org
(281) 583-4087

DASMA Door and Access Systems Manufacturers Association, International
(Formerly: National Association of Garage Door Manufacturers)
c/o Thomas Associates, Inc.
1300 Sumner Ave.
Cleveland, OH 44115-2851
www.taol.com/dasma
(216) 241-7333

DHI Door and Hardware Institute
(Formerly: National Builders Hardware Association)
14170 Newbrook Dr.
Chantilly, VA 20151-2223
www.dhi.org
(703) 222-2010

DIPRA Ductile Iron Pipe Research Association
245 Riverchase Pkwy East, Suite O
Birmingham, AL 35244
www.dipra.org
(205) 402-8702

DLPA Decorative Laminate Products Association
(Dissolved in 1995—Now part of KCMA.)

ECSA Exchange Carriers Standards Association
(See ATIS)

EIA Electronic Industries Association
2500 Wilson Blvd.
Arlington, VA 22201
www.eia.org
(703) 907-7500

EIMA EIFS Industry Members Association
3000 Corporate Center Dr., Suite 270
Morrow, GA 30260-4116
www.eifsfacts.com
(800) 294-3462
(770) 968-7945

EJMA Expansion Joint Manufacturers Association
25 N. Broadway
Tarrytown, NY 10591-3201
www.ejma.org
(914) 332-0040

ETL ETL Testing Laboratories, Inc.
(Now part of ITS)

FCI Fluid Controls Institute
c/o Thomas Associates, Inc.
1300 Sumner Ave.
Cleveland, OH 44115-2851
www.taol.com/fci
(216) 241-7333

FCICA Floor Covering Installation Contractors Association
P.O. Box 948
Dalton, GA 30722-0948
(706) 226-5488

FGMA Flat Glass Marketing Association
(See GANA)

FM Factory Mutual System
1151 Boston-Providence Tnpk.
P.O. Box 9102
Norwood, MA 02062-9102
www.factorymutual.com
(781) 762-4300

GA Gypsum Association
810 First St., NE, Suite 510
Washington, DC 20002
www.usg.com
(202) 289-5440

GANA Glass Association of North America
(Formerly: Flat Glass Marketing Association)
2945 Southwest Wanamaker Dr., Suite A
Topeka, KS 66614
www.glasswebsite.com/gana
(913) 266-7013

GRI Geosynthetic Research Institute
475 Kendron Ave.
Folsom, PA 19033
www.drexel.edu/gri
(610) 522-8440

HEI Heat Exchange Institute
c/o Thomas Associates, Inc.
1300 Sumner Ave.
Cleveland, OH 44115-2851
www.taol.com/hei
(216) 241-7333

HI Hydraulic Institute
9 Sylvan Way
Parsippany, NJ 07054-3802
(888) 786-7744
(973) 267-9700

HI Hydronics Institute
Division of Gas Appliance
Manufacturers Association
P.O. Box 218
35 Russo Pl.
Berkeley Heights, NJ 07922
www.gamanet.org
(908) 464-8200

HMA Hardwood Manufacturers Association
(Formerly: Southern Hardwood Lumber Manufacturers Association)
400 Penn Center Blvd., Suite 530
Pittsburgh, PA 15235-5605
www.hardwood.org
(412) 829-0770

HPVA Hardwood Plywood and Veneer Association
1825 Michael Farraday Dr.
P.O. Box 2789
Reston, VA 22195-0789
www.hpva.org
(703) 435-2900

IAS International Approval Services
Division of Canadian Standards Association
8501 East Pleasant Valley Rd.
Cleveland, OH 44131
www.iasapprovals.org
(216) 524-4990

IBD Institute of Business Designers
(Now part of IIDA)

ICC International Code Council
5203 Leesburg Pike #708
Falls Church, VA 22041
ww.intlcode.org
(703) 931-4533

ICEA Insulated Cable Engineers Association
P.O. Box 440
South Yarmouth, MA 02664
www.icea.net
(508) 394-4424

IEC International Electrotechnical Commission
(Available from ANSI)
11 West 42nd St., 13th Floor
New York, NY 10036-8002
www.ansi.org
(888) 267-4783
(212) 642-4900

IEEE Institute of Electrical and Electronics Engineers
345 E. 47th St.
New York, NY 10017-2394
www.ieee.org
(800) 678-4333
(212) 705-7900

IESNA Illuminating Engineering Society of North America
120 Wall St., 17th Floor
New York, NY 10005-4001
www.iesna.org
(212) 248-5000

IGCC Insulating Glass Certification Council
P.O. Box 9
Henderson Harbor, NY 13651
(315) 938-7444

IIDA International Interior Design Association
341 Merchandise Mart
Chicago, IL 60654-1104
www.iida.com
(800) 888-4432
(312) 467-1950

ILI Indiana Limestone Institute of America
Stone City Bank Building, Suite 400
Bedford, IN 47421
www.iliai.com
(812) 275-4426

IMSA International Municipal Signal Association
P.O. Box 539
165 E. Union St.
Newark, NY 14513
www.imsasafety.org
(800) 723-4672
(315) 331-2182

INCE Institute of Noise Control Engineering
P.O. Box 3206, Arlington Branch
Poughkeepsie, NY 12603
www.ince.org
(914) 462-4006

IRI HSB Industrial Risk Insurers
P.O. Box 5010
85 Woodland St.
Hartford, CT 06102-5010
www.industrailrisk.com
(800) 520-7300
(860) 520-7300

ISA — ISA—International Society for Measurement and Control
P.O. Box 12277
67 Alexander Dr.
Research Triangle Park, NC 27709
www.isa.org
(919) 549-8411

ISEA — Industrial Safety Equipment Association
1901 N. Moore St., Suite 808
Arlington, VA 22209
www.safetycentral.org/isea
(703) 525-1695

ISS — Iron and Steel Society
410 Commonwealth Dr.
Warrendale, PA 15086-7512
www.issource.org
(412) 776-1535

ITS — Intertek Testing Services
(Formerly: Inchcape Testing Services)
P.O. Box 2040
3933 US Route 11
Cortland, NY 13045-7902
www.itsglobal.com
(800) 345-3851
(607) 753-6711

KCMA — Kitchen Cabinet Manufacturers Association
(Formerly: National Kitchen Cabinet Association)
1899 Preston White Dr.
Reston, VA 22191
www.kcma.org
(703) 264-1690

LGSI — Light Gage Structural Institute
P.O. Box 560746
The Colony, TX 75056
(972) 625-4560

LIA — Lead Industries Association, Inc.
295 Madison Ave., Suite 808
New York, NY 10017
www.leadinfo.com
(800) 422-5323
(212) 578-4750

LMA — Laminating Materials Association
(Formerly: American Laminators Association)
116 Lawrence St.
Hillsdale, NJ 07642-2730
www.lma.org
(201) 664-2700

LPI — Lightning Protection Institute
3335 N. Arlington Heights Rd., Suite E
Arlington Heights, IL 60004-7700
www.lightning.org
(800) 488-6864
(847) 577-7200

MBMA — Metal Building Manufacturers Association
c/o Thomas Associates, Inc.
1300 Sumner Ave.
Cleveland, OH 44115-2851
www.taol.com/mbma
(216) 241-7333

MCAA — Mechanical Contractors Association of America
1385 Piccard Dr.
Rockville, MD 20850-4329
www.mcaa.org
(301) 869-5800

MFMA — Maple Flooring Manufacturers Association
(Formerly: Wood and Synthetic Flooring Institute)
60 Revere Dr., Suite 500
Northbrook, IL 60062
www.maplefloor.com
(847) 480-9138

MFMA — Metal Framing Manufacturers Association
401 N. Michigan Ave.
Chicago, IL 60611
(312) 644-6610

MHIA — Material Handling Industry Association
8720 Red Oak Blvd., Suite 201
Charlotte, NC 28217-3957
www.mhia.org
(800) 345-1815
(704) 676-1190

MIA — Marble Institute of America
30 Eden Alley, Suite 301
Columbus, OH 43215
www.marble-institute.com
(614) 228-6194

MIA — Masonry Institute of America
2550 Beverly Blvd.
Los Angeles, CA 90057
www.masonryinstitute.org
(213) 388-0472

ML/SFA Metal Lath/Steel Framing Association
8 South Michigan Ave., Suite 1000
Chicago, IL 60603
(312) 456-5590

MRCA Midwest Roofing Contractors Association
4840 W. 15th St., Suite 1000
Lawrence, KS 66049
www.mrca.org
(913) 843-4888

MSS Manufacturers Standardization Society of the Valve and Fittings Industry
127 Park St., NE
Vienna, VA 22180-4602
www.mss-hq.com
(703) 281-6613

NAA National Arborist Association
P.O. Box 1094
Amherst, NH 03031-1094
www.natlarb.com
(800) 733-2622
(603) 673-3311

NAAMM National Association of Architectural Metal Manufacturers
8 South Michigan Ave., Suite 1000
Chicago, IL 60603
www.gss.net/naamm
(312) 322-0405

NAAMM North American Association of Mirror Manufacturers
(See GANA)
2945 Southwest Wanamaker Dr., Suite A
Topeka, KS 66614
www.glasswebsite.com/naamm
(913) 266-7013

NACE NACE International
(Formerly: National Association of Corrosion Engineers)
P.O. Box 218340
Houston, TX 77218-8340
(281) 492-0535
(281) 492-8254

NAGDM National Association of Garage Door Manufacturers
(See DASMA)

NAIMA North American Insulation Manufacturers Association
(Formerly: Thermal Insulation Manufacturers Association)
44 Canal Center Plaza, Suite 310
Alexandria, VA 22314
www.naima.org
(703) 684-0084

NAMI National Accreditation & Management Institute, Inc.
P.O. Box 366
207 S. Washington St.
Berkeley Springs, WV 25411
(304) 258-5100

NAPA National Asphalt Pavement Association
NAPA Building
5100 Forbes Blvd.
Lanham, MD 20706-4413
www.hotmix.org
(888) 468-6499
(301) 731-4748

NBHA National Builders Hardware Association
(See DHI)

NBGQA National Building Granite Quarries Association, Inc
1220 L. St., NW, Suite 100-167
Washington, DC 20005
(800) 557-2848

NCAC National Council of Acoustical Consultants
66 Morris Ave., Suite 1A
Springfield, NJ 07081
www.ncac.com
(973) 564-5859

NCCA National Coil Coaters Association
401 N. Michigan Ave.
Chicago, IL 60611
www.coilcoaters.org
(312) 321-6894

NCMA National Concrete Masonry Association
2302 Horse Pen Rd.
Herndon, VA 20171-3499
www.ncma.org
(703) 713-1900

NCPI National Clay Pipe Institute
P.O. Box 759
253-80 Center St.
Lake Geneva, WI 53147
www.ncpi.org
(414) 248-9094

NCRPM	National Council on Radiation Protection and Measurements 7910 Woodmont Ave., Suite 800 Bethesda, MD 20814-3095 www.ncrp.com (800) 229-2652 (301) 657-2652
NCSPA	National Corrugated Steel Pipe Association 1255 23rd St., NW, Suite 850 Washington, DC 20037 www.ncspa.org (202) 452-1700
NEBB	Natural Environmental Balancing Bureau 8575 Grovemont Circle Gaithersburg, MD 20877-4121 www.mcaa.org/nebb.htm www.nebb.org (301) 977-3698
NECA	National Electrical Contractors Association 3 Bethesda Metro Center, Suite 1100 Bethesda, MD 20814-3299 www.necanet.org (301) 657-3110
NEI	National Elevator Industry 185 Bridge Plaza North, Suite 310 Fort Lee, NJ 07024 (201) 944-3211
NELMA	Northeastern Lumber Manufacturers Association 272 Tuttle Rd. P.O. Box 87A Cumberland Center, ME 04021 www.nelma.org (207) 829-6901
NEMA	National Electrical Manufacturers Association 1300 N 17th St., Suite 1847 Rosslyn, VA 22209 www.nema.org (703) 841-3200
NETA	InterNational Electrical Testing Association P.O. Box 687 106 Stone St. Morrison, CO 80465-1526 www.electricnet.com/neta (303) 697-8441
NFPA	National Fire Protection Association One Batterymarch Park P.O. Box 9101 Quincy, MA 02269-9101 www.nfpa.org (800) 344-3555 (617) 770-3000
NFPA	National Forest Products Association (See AFPA)
NFRC	National Fenestration Rating Council Incorporated 1300 Spring St., Suite 500 Silver Spring, MD 20910 www.nfrc.org (301) 589-6372
NGA	National Glass Association 8200 Greensboro Drive, 3rd Floor McLean, VA 22102-3881 (703) 442-4890
NHLA	National Hardwood Lumber Association P.O. Box 34518 Memphis, TN 38184-0518 www.natlhardwood.org (901) 377-1818
NIA	National Insulation Association (Formerly: National Insulation and Abatement Contractors Association) 99 Canal Center Plaza, Suite 222 Alexandria, VA 22314 www.insulation,org (703) 683-6422
NIAC	National Insulation and Abatement Contractors Association (See NIA)
NKCA	National Kitchen Cabinet Association (See KCMA)
NLGA	National Lumber Grades Authority #406-First Capital Pl. 960 Quayside Dr. New Westminster, BC V3M 6G2 CANADA (604) 524-2393
NOFMA	National Oak Flooring Manufacturers Association P.O. Box 3009 Memphis, TN 38173-0009 www.nofma.org (901) 526-5016

NPA	National Parking Association 1112 16th Street, NW, Suite 300 Washington, DC 20036 (202) 296-4336 (800) 647-7275
NPCA	National Paint and Coatings Association 1500 Rhode Island Ave., NW Washington, DC 20005-5597 www.paint.org (202) 462-6272
NRCA	National Roofing Contractors Association O'Hare International Center 10255 W. Higgins Rd., Suite 600 Rosemont, IL 60018-5607 www.roofonline.org (800) 323-9545 (847) 299-9070
NRMCA	National Ready Mixed Concrete Association 900 Spring St. Silver Spring, MD 20910 www.nrmca.org (301) 587-1400
NSA	National Stone Association 1415 Elliot Pl., NW Washington, DC 20007 www.aggregates.org (800) 342-1415 (202) 342-1100
NSF	NSF International (Formerly: National Sanitation Foundation) 3475 Plymouth Rd. Ann Arbor, MI 48105 www.nsf.org (734) 769-8010
NSSEA	National School Supply and Equipment Association 8300 Colesville Rd., Suite 250 Silver Spring, MD 20910 www.nssea.org (800) 395-5550 (301) 495-0240
NTMA	National Terrazzo and Mosaic Association 110 E. Market St., Suite 200-A Leesburg, VA 20176-3122 www.ntma.com (800) 323-9736 (703) 779-1022
NUSIG	National Uniform Seismic Installation Guidelines P.O. Box 0933 Alamo, CA 94507 (925) 555-6331
NWMA	National Woodwork Manufacturers Association (See NWWDA)
NWWDA	National Wood Window and Door Association (Formerly: National Woodwork Manufacturers Association) 1400 E. Touhy Ave. Des Plaines, IL 60018 www.nwwda.org (800) 223-2301 (847) 299-5200
PATMI	Powder Actuated Tool Manufacturers' Institute 1603 Boonslick Rd. St. Charles, MO 63301-2244 (314) 947-6610
PCA	Portland Cement Association 5420 Old Orchard Rd. Skokie, IL 60077-1083 www.portcement.org (847) 966-6200
PCI	Precast/Prestressed Concrete Institute 175 W. Jackson Blvd. Chicago, IL 60604 www.pci.org (312) 786-0300
PDCA	Painting and Decorating Contractors of America 3913 Old Lee Hwy, Suite 33-B Fairfax, VA 22030 www.pdca.com (800) 332-7322 (703) 359-0826
PDI	Plumbing and Drainage Institute 45 Bristol Dr., Suite 101 South Easton, MA 02375 www.pdionline.org (800) 589-8956 (508) 230-3516
PEI	Porcelain Enamel Institute 4004 Hilsboro Pike, Suite 224-B Nashville, TN 37215 www.porcelainenamel.com (615) 385-5357

PGI — PVC Geomembrane Institute/Technology Program
University of Illinois-Urbana Champaign
205 N. Mathews Ave.
2215 Newmark Civil Engineering Lab
Urbana, IL 61801
pgi-tp.ce.uiuc.edu
(217) 333-3929

PIMA — Photographic and Imaging Manufacturers Association
550 Mamaroneck Ave., Suite 307
Harrison, NY 10528
www.pima.net
(914) 698-7603

PPFA — Plastics Pipe and Fittings Association
800 Roosevelt Rd., Building C, Suite 20
Glen Ellyn, IL 60137-5833
(888) 314-6774
(630) 858-6540

PPI — Plastics Pipe Institute
(The Society of the Plastics Industry, Inc.)
1801 K St., NW, Suite 600K
Washington, DC 20006
www.plasticpipe.org
(202) 974-5306

RCMA — Roof Coatings Manufacturers Association
Center Park
4041 Powder Mill Rd., Suite 404
Calverton, MD 20705
www.roofcoatings.org
(301) 348-2003

RCSC — Research Council on Structural Connections
Sargent & Lundy
55 E. Monroe St.
Chicago, IL 60603
(312) 269-2424

RFCI — Resilient Floor Covering Institute
966 Hungerford Dr., Suite 12-B
Rockville, MD 20850-1714
(301) 340-8580

RMA — Rubber Manufacturers Association
1400 K St., NW, Suite 900
Washington, DC 20005
www.rma.org
(800) 220-7620
(202) 682-4800

SAE — SAE International
400 Commonwealth Dr.
Warrendale, PA 15096-0001
For publications: Call (724) 776-4970
www.sae.org
(724) 776-4841

SDI — Steel Deck Institute
P.O. Box 25
Fox River Grove, IL 60021
www.sdi.org
(847) 462-1930

SDI — Steel Door Institute
30200 Detroit Rd.
Cleveland, OH 44145-1967
www.steeldoor.org
(440) 899-0010

SEFA — Scientific Equipment and Furniture Association
7 Wildbird Lane
Hilton Head Island, SC 29926
www.sefalabfurn.com
(843) 689-6878

SEGD — Society for Environmental Graphic Design
401 F St., NW, Suite 333
Washington, DC 20001-2728
(202) 683-5555

SGCC — Safety Glazing Certification Council
P.O. Box 9
Henderson Harbor, NY 13651
(315) 938-7444

SHLMA — Southern Hardwood Lumber Manufacturers Association
(See HMA)

SIGMA — Sealed Insulating Glass Manufacturers Association
401 N. Michigan Ave.
Chicago, IL 60611-4267
www.sigmaonline.org/sigma
(312) 644-6610 x3279

SJI — Steel Joist Institute
3127 10th Ave., North Ext.
Myrtle Beach, SC 29577-6760
(803) 626-1995

SMA — Screen Manufacturers Association
2850 S. Ocean Blvd., Suite 114
Palm Beach, FL 33480-5535
(561) 533-0991

SMACNA	Sheet Metal and Air Conditioning Contractors' National Association 4201 Lafayette Center Dr. P.O. Box 221230 Chantilly, VA 20151-1209 www.smacna.org (703) 803-2980
SPI	The Society of the Plastics Industry, Inc. Spray Polyurethane Division 1801 K St., NW, Suite 600K Washington, DC 20006 www.socplas.org (800) 951-2001 (202) 974-5200
SPIB	Southern Pine Inspection Bureau 4709 Scenic Hwy Pensacola, FL 32504-9094 www.spib.org (850) 434-2611
SPRI	SPRI (Formerly: Single Ply Roofing Institute) 200 Reservoir St., Suite 309A Needham, MA 02494-3034 www.spri.org (781) 444-0242
SSINA	Speciality Steel Industry of North America c/o Collier, Shannon Rill & Scott 3050 K St., NW, Suite 400 Washington, DC 20007 www.ssina.com (800) 982-0355 (202) 342-8630
SSPC	SSPC: The Society for Protective Coatings 40 24th St., 6th Floor Pittsburgh, PA 15222-4656 www.sspc.org (800) 837-8303 (412) 281-2331
SSPMA	Sump and Sewage Pump Manufacturers Association P.O. Box 647 Northbrook, IL 60065-0647 (847) 559-9233
STI	Steel Tank Institute 570 Oakwood Rd. Lake Zurich, IL 60047-1559 www.steeltank.com (847) 438-8265
SWI	Steel Window Institute c/o Thomas Associates, Inc. 1300 Sumner Ave. Cleveland, OH 44115-2851 www.taol.com/swi (216) 241-7333
SWPA	Submersible Wastewater Pump Association 1806 Johns Dr. Glenview, IL 60025-1657 (847) 729-7972
SWRI	Sealant, Waterproofing and Restoration Institute 2841 Main St. Kansas City, MO 64108 www.swrionline.org (816) 472-7974
TCA	Tile Council of America P.O. Box 1787 Clemson, SC 29633 www.tileusa.com (864) 646-8453
TIMA	Thermal Insulation Manufacturers Association (See NAIMA)
TPI	Truss Plate Institute 583 D'Onofrio Dr., Suite 200 Madison, WI 53719 (608) 833-5900
TPI	Turfgrass Producers International (Formerly: American Sod Producers Association) 1855-A Hicks Rd. Rolling Meadows, IL 60008 www.turfgrassod.org (800) 405-8873 (847) 705-9898
UFAC	Upholstered Furniture Action Council P.O. Box 2436 High Point, NC 27261 (910) 885-6085
UL	Underwriters Laboratories Inc. 333 Pfingsten Rd. Northbrook, IL 60062 www.ul.com (800) 704-4050 (847) 272-8800

UNI	Uni-Bell PVC Pipe Association 2655 Villa Creek Dr., Suite 155 Dallas, TX 75234 members.aol.com/unibell (972) 243-3902
USITT	USITT: The American Association of Design and Production Professionals in the Performing Arts 6443 Ridings Rd. Syracuse, NY 13206-1111 www.culturenet.ca/usitt (800) 938-7488 (315) 463-6463
USP	U.S. Pharmacopoeia (Formerly: U.S. Pharmacopoeial Convention) 12601 Twinbrook Pkwy. Rockville, MD 20852-1790 www.usp.org (800) 277-8772 (301) 881-0666
WA	Wallcoverings Association 401 N. Michigan Ave. Chicago, IL 60611-4267 (312) 644-6610
WASTEC	Waste Equipment Technology Association 4301 Connecticut Ave., NW, Suite 300 Washington, DC 20008 (202) 244-4700
WCLIB	West Coast Lumber Inspection Bureau P.O. Box 23145 Portland, OR 97281-3145 (800) 283-1486 (503) 639-0651
WCMA	Window Covering Manufacturers Association (Formerly: American Window Covering Manufacturers Association) 355 Lexington Ave., 17th Floor New York, NY 10017-6603 (212) 661-4261
WEF	Water Environment Federation (Formerly: Water Pollution Control Federation) 601 Wythe St. Alexandria, VA 22314-1994 www.wef.org (800) 666-0206 (703) 684-2400
WIC	Woodwork Institute of California P.O. Box 980247 West Sacramento, CA 95798-0247 www.wicnet.org (916) 372-9943
WMMPA	Wood Moulding & Millwork Producers Association 507 First St. Woodland, CA 95695 www.wmmpa.com (800) 550-7889 (530) 661-9591
WPCF	Water Pollution Control Federation (See WEF)
WRI	Wire Reinforcement Institute 301 E. Sandusky St. Findlay, OH 45840 www.bright.net/~rreiter (419) 425-9473
WSC	Water Systems Council Building C, Suite 20 800 Roosevelt Rd. Glen Ellyn, IL 60137 (630) 545-1762
WSFI	Wood and Synthetic Flooring Institute (See MFMA)
WWPA	Western Wood Products Association Yeon Building 522 SW 5th Ave. Portland,OR 97204-2122 www.wwpa.org (503) 224-3930

UNIFIED SOIL CLASSIFICATION

MAJOR DIVISIONS		SYMBOL	TYPICAL NAMES
GRAVELS (More than half of coarse fraction larger than #4 sieve size)	Clean gravel	**GW**	Well-graded gravel, gravel-sand mixtures
		GP	Poorly graded gravel, gravel-sand mixtures
	Gravel with fines	**GM**	Silty gravels, gravel-sand-silt mixtures
		GC	Clayey gravels, gravel-sand clay mixtures
SANDS (More than half of coarse fraction smaller than #4 sieve size)	Clean Sands	**SW**	Well-graded sands, gravelly sands
		SP	Poorly graded sands, gravelly sands
	Sands with fines	**SM**	Silty sands, sand-silt mixture
		SC	Clayey sands, sand-clay mixtures
FINE-GRAINED SOILS (More than half of material is smaller than #200 sieve)	Silts & Clays (LL < 50)	**ML**	Inorganic silt & very fine sands, clayey fine sands/silts with slight plasticity
		CL	Inorganic clays of low to medium plasticity, sandy/gravelly clays
		OL	Organic silts and organic silty clays of low plasticity
	Silts & Clays (LL > 50)	**MH**	Inorganic silts, fine sandy/silty soils, elastic silts
		CH	Inorganic clays, medium to high plasticity, fat clays
	Highly organic soils	**PT**	Peat and highly organic soils

Definition of Soil by Grain size

Sieve size	Corresponding soil classification
12" (304.8 mm) or more	Boulders
3" (76.2 mm) to 12" (304.8 mm)	Cobbles
¾" (19.05 mm) to 3" (76.2 mm)	Coarse gravel
No. 4 to ¾" (19.05 mm)	Fine gravel
No. 4 to No. 10	Coarse sand
No. 10 to No. 40	Medium sand
No. 40 to No. 200	Fine sand
Passing through No. 200	Silt and clay fines

U.S.A. /Metric Sieve Sizes

This chart shows the various sieve-size openings and their metric conversions.

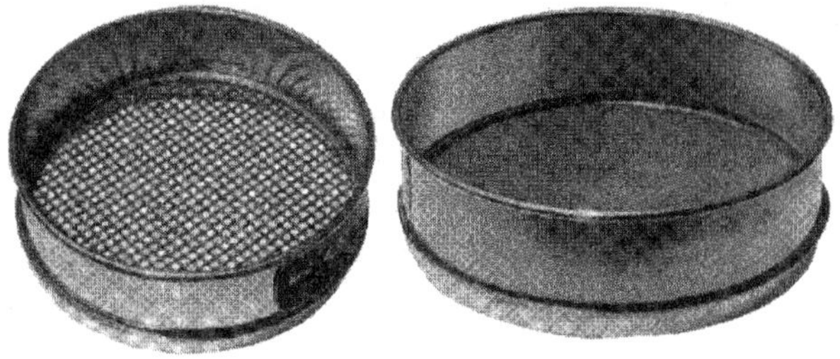

U.S.A. Sieve Series and Equivalents—A.S.T.M. E-11-87					
Sieve Designation		Sieve Opening		Nominal Wire Diameter	
Standard (a)	Alternative	mm	in (approx. equivnts.)	mm	in (approx. equivnts.)
125 mm	5"	125	5.00"	8.00	.3150"
106 mm	4.24"	106	4.24"	6.40	.2520"
100 mm	4"(b)	100	4.00"	6.30	.2480"
90 mm	3.5"	90	3.50"	6.08	.2394"
75 mm	3"	75	3.00"	5.80	.2283"
63 mm	2.5"	63	2.50"	5.50	.2165"
53 mm	2.12"	53	2.12"	5.15	.2028"
50 mm	2"(b)	50	2.00"	5.05	.1988"
45 mm	1.75"	45	1.75"	4.85	.1909"
37.5 mm	1.5"	37.5	1.50"	4.59	.1807"
31.5 mm	1.25"	31.5	1.25"	4.23	.1665"
26.5 mm	1.06"	26.5	1.06"	3.90	.1535"
25.0 mm	1"(b)	25.0	1.00"	3.80	.1496"
22.4 mm	7/8"	22.4	0.875"	3.50	.1378"
19.0 mm	3/4"	19.0	0.750"	3.30	.1299"
16.0 mm	5/8"	16.0	0.625"	3.00	.1181"
13.2 mm	.530"	13.2	0.530"	2.75	.1083"
12.5 mm	1/2"(b)	12.5	0.500"	2.67	.1051"
11.2 mm	7/16"	11.2	0.438"	2.45	.0965"
9.5 mm	3/8"	9.5	0.375"	2.27	.0894"
8.0 mm	5/16"	8.0	0.312"	2.07	.0815"
6.7 mm	.265"	6.7	0.265"	1.87	.0736"
6.3 mm	1/4"(b)	6.3	0.250"	1.82	.0717"
5.6 mm	No. 3-1/2(c)	5.6	0.223"	1.68	.0661"
4.75 mm	No. 4	4.75	0.187"	1.54	.0606"
4.00 mm	No. 5	4.00	0.157"	1.37	.0539"
3.35 mm	No. 6	3.35	0.132"	1.23	.0484"
2.80 mm	No. 7	2.80	0.11"	1.10	.0430"
2.36 mm	No. 8	2.36	0.0937"	1.00	.0394"
2.00 mm	No. 10	2.00	0.0787"	.900	.0345"
1.70 mm	No. 12	1.70	0.0661"	.810	.0319"
1.40 mm	No. 14	1.40	0.0555"	.725	.0285"
1.18 mm	No. 16	1.18	0.0469"	.650	.0256"
1.00 mm	No. 18	1.00	0.0394"	.580	.0228"
850 µm	No. 20	0.850	0.0331"	.510	.0201"
710 µm	No. 25	0.710	0.0278"	.450	.0177"
660 µm	No. 30	0.600	0.0234"	.390	.0154"
500 µm	No. 35	0.500	0.0197"	.340	.0134"
425 µm	No. 40	0.425	0.0165"	.290	.0114"
355 µm	No. 45	0.355	0.0139"	.247	.0097"
300 µm	No. 50	0.300	0.0117"	.215	.0085"
250 µm	No. 60	0.250	0.0098"	.180	.0071"
212 µm	No. 70	0.212	0.0083"	.152	.0060"
180 µm	No. 80	0.180	0.0070"	.131	.0052"
150 µm	No. 100	0.150	0.0059"	.110	.0043"
125 µm	No. 120	0.125	0.0049"	.091	.0036"
106 µm	No. 140	0.106	0.0041"	.076	.0030"
90 µm	No. 170	0.090	0.0035"	.064	.0025"
75 µm	No. 200	0.075	0.0029"	.053	.0021"
63 µm	No. 230	0.063	0.0025"	.044	.0017"
53 µm	No. 270	0.053	0.0021"	.037	.0015"
45 µm	No. 325	0.045	0.0017"	.030	.0012"
38 µm	No. 400	0.038	0.0015"	.025	.0010"
32 µm	No. 450		0.00126"	.0011	
25 µm	No. 500		0.00098"	.001	
20 µm	No. 635		0.00079"	.0008	

(a) These standard designations correspond to the values for test sieve apertures recommended by the International Standards Organization Geneva, Switzerland.
(b) These sieves are not in the fourth root of 2 Series, but they have been included because they are in common usage.
(c) These numbers (3-1/2 to 400) are the approximate number of openings per linear inch but it is preferred that the sieve be identified by the standard designation in millimeters or microns (1000 microns = 1 mm.)

Mechanical Properties of Various Rocks

MECHANICAL PROPERTIES OF VARIOUS ROCKS					
Rock	**Young's Modulus at Zero Load** (10^5 kg/cm^2)	**Bulk Density** (g/cm^3)	**Porosity** (percent)	**Compressive Strength** (kg/cm^2)	**Tensile Strength** (kg/cm^2)
Granite	2 - 6	2.6-2.7	0.5-1.5	1,000-2,500	70-250
Microgranite	3 - 8				
Syenite	6 - 8				
Diorite	7-10			1,800-3,000	150-300
Dolerite	8-11	3.0-3.05	0.1-0.5	2,000-3,500	150-350
Gabbro	7-11	3.0-3.1	0.1-0.2	1,000-3,000	150-300
Basalt	6-10	2.8-2.9	0.1-1.0	1,500-3,00	100-300
Sandstone	0.5-8	2.0-2.6	5 - 25	200-1,700	40-250
Shale	1-3.5	2.0-2.4	10 - 30	100-1,000	20-100
Mudstone	2 - 5				
Limestone	1 - 8	2.2-2.6	5 - 20	300-3,500	50-250
Dolomite	4-8.4	2.5-2.6	1 - 5	800-2,500	150-250
Coal	1 - 2			50-500	20-50
Quartzite		2.65	0.1-0.5	1,500-3,000	100-300
Gneiss		2.9-3.0	0.5-1.5	500-2,000	50-200
Marble		2.6-2.7	0.5-2	1,000-2,500	70-200
Slate		2.6-2.7	0.1-0.5	1,000-2,000	70-200

Note:
1. For the igneous rocks listed above Poisson's ratio is approximately 0.25.
2. For a certain rock type, the strength normally increases with increase in density and increase in Young's modulus. (After Farmer, 1968)
3. Taken from *"Foundation Engineering Handbook"* by Winterkorn and Fong, Van Nostrand Reinhold, pg. 72.

Nonreinforced Culvert Pipe—Sizes and Weights

Dimensions and Approximate Weights of Circular Concrete Pipe, Bell and Spigot Joint.

ASTM C 14–Nonreinforced Sewer and Culvert Pipe, Bell and Spigot Joint						
Internal Diameter, inches	Class 1		Class 2		Class 3	
	Minimum Wall Thickness, inches	Average Weight, pounds per foot	Minimum Wall Thickness, inches	Average Weight, pounds per foot	Minimum Wall Thickness, inches	Average Weight, pounds per foot
4	5/8	9.5	3/4	13	3/4	13
6	5/8	17	3/4	20	7/8	21
8	3/4	27	7/8	31	1-1/8	36
10	7/8	37	1	42	1-1/4	50
12	1	50	1-3/8	68	1-3/4	90
15	1-1/4	78	1-5/8	100	1-7/8	120
18	1-1/2	105	2	155	2-1/4	165
21	1-3/4	159	2-1/4	205	2-3/4	260
24	2-1/8	200	3	315	3-3/8	350
27	3-1/4	390	3-3/4	450	3-3/4	450
30	3-1/2	450	4-1/4	540	4-1/4	540
33	3-3/4	520	4-1/2	620	4-1/2	620
36	4	580	4-3/4	700	4-3/4	700

Reinforced Concrete Pipe—Sizes and Weights

ASTM C 76–Reinforced Concrete Culvert, Storm Drain and Sewer Pipe, Bell and Spigot Joint				
Internal Diameter, inches	Wall A		Wall B	
	Minimum Wall Thickness, inches	Average Weight, pounds per foot	Minimum Wall Thickness, inches	Average Weight, pounds per foot
12	1-3/4	90	2	106
15	1-7/8	120	2-1/4	148
18	2	155	2-1/2	200
21	2-1/4	205	2-3/4	260
24	2-1/2	265	3	325
27	2-5/8	310	3-1/4	388
30	2-3/4	363	3-1/2	459

These tables are based on concrete weighing 150 pounds per cubic foot and will vary with heavier or lighter weight concrete.

(By permission: American Concrete Pipe Association, Irving, Texas)

Approximate Weight per Foot/Corrugated Steel Pipe

(Estimated Average Weights—Not for Specification Use)

1½" x ¼" Corrugation

Inside Diameter, in.	Specified Thickness, in.	Galvanized & ALUMINIZED	Full Coated
6	0.052	4	5
	0.064	5	6
8	0.052	5	6
	0.064	6	7
10	0.052	6	7
	0.064	7	8

2⅔" x ½" Corrugation

Inside Diameter, in.	Specified Thickness, in.	Galvanized & ALUMI-NIZED*	Full Coated	Coated & PAVED-INVERT	SMOOTH-FLO	HEL-COR CL
12	0.052	8	10	13		
	0.064	10	12	15		
	0.079	12	14	17		
15	0.052	10	13	16	26	
	0.064	12	15	18	28	
	0.079	15	18	21	31	
18	0.052	12	16	19	31	
	0.064	15	19	22	34	
	0.079	18	22	25	37	
21	0.052	14	18	23	36	
	0.064	17	21	26	39	
	0.079	21	25	30	43	
24	0.052	15	20	26	41	
	0.064	19	24	30	45	65
	0.079	24	29	35	50	69
	0.109	33	38	44	59	77
30	0.052	20	26	32	51	
	0.064	24	30	36	55	82
	0.079	30	36	42	60	87
	0.109	41	47	53	72	96
36	0.052	24	31	39	50	
	0.064	29	36	44	65	98
	0.079	36	43	51	75	104
	0.109	49	56	64	90	116
	0.138	62	69	77	100	127
42	0.052	28	36	45	71	
	0.064	34	42	51	77	114
	0.079	42	50	59	85	121
	0.109	57	65	74	100	135
	0.138	72	80	89	115	149
48	0.064	38	48	57	85	128
	0.079	48	58	67	95	138
	0.109	65	75	84	112	154
	0.138	82	92	101	129	170
	0.168	100	110	119	147	186
54	0.079	54	65	76	105	156
	0.109	73	84	95	124	173
	0.138	92	103	114	143	191
	0.168	112	123	134	163	209
60	0.109	81	92	106	140	192
	0.138	103	114	128	162	212
	0.168	124	135	149	183	232
66	0.109	89	101	117	160	211
	0.138	113	125	141	180	233
	0.168	137	149	165	210	255
72	0.138	123	137	154	210	254
	0.168	149	163	180	236	278
78	0.168	161	177	194	260	302
84	0.168	173	190	208	270	325

3" x 1" or 5" x 1" Corrugation

Inside Diameter, in.	Specified Thickness, in.	Galvanized & ALUMI-NIZED*	Full Coated	Coated & PAVED-INVERT	SMOOTH-FLO	HEL-COR CL
54	0.064	50	66	84	138	197
	0.079	61	77	95	149	207
	0.109	83	100	118	171	226
	0.138	106	123	140	194	245
	0.168	129	146	163	217	264
60	0.064	55	73	93	153	218
	0.079	67	86	105	165	229
	0.109	92	110	130	190	251
	0.138	118	136	156	216	272
	0.168	143	161	181	241	293
66	0.064	60	80	102	168	240
	0.079	74	94	116	181	252
	0.109	101	121	143	208	276
	0.138	129	149	171	236	299
	0.168	157	177	199	264	322
72	0.064	66	88	111	183	262
	0.079	81	102	126	197	275
	0.109	110	132	156	227	301
	0.138	140	162	186	257	326
	0.168	171	193	217	288	351
78	0.064	71	95	121	198	
	0.079	87	111	137	214	298
	0.109	119	143	169	246	326
	0.138	152	176	202	279	353
	0.168	185	209	235	312	380
84	0.064	77	102	130	213	
	0.079	94	119	147	230	321
	0.109	128	154	182	264	351
	0.138	164	189	217	300	379
	0.168	199	224	253	335	409
90	0.064	82	109	140	228	
	0.079	100	127	158	246	
	0.109	137	164	195	283	376
	0.138	175	202	233	321	406
	0.168	213	240	271	359	438
96	0.064	87	116	149	242	
	0.079	107	136	169	262	
	0.109	147	176	209	302	401
	0.138	188	217	250	343	433
	0.168	228	257	290	383	467
102	0.064	93	124	158	258	
	0.079	114	145	179	279	
	0.109	155	186	220	320	426
	0.138	198	229	263	363	460
	0.168	241	272	306	406	496
108	0.079	120	153	188	295	
	0.109	165	198	233	340	
	0.138	211	244	279	386	487
	0.168	256	289	324	431	525
114	0.079	127	162	199	312	
	0.109	174	209	246	359	
	0.138	222	257	294	407	514
	0.168	271	306	343	456	554
120	0.109	183	220	259	378	
	0.138	234	271	310	429	541
	0.168	284	321	360	479	583
126	0.138	247	285	326	452	
132	0.138	259	299	342	474	
	0.168	314	354	397	529	
138	0.138	270	312	357	495	
	0.168	328	370	415	553	
144	0.168	344	388	435	579	

* Weights for polymer-coated pipe are 1% to 4% higher, varying by gage.

(By permission: CONTECH Construction Products Inc., Middletown, Ohio)

Ductile Iron Specifications, 3- to 36-Inch Push-on Joint Pipe

Standard Dimensions and Weights of
3 in through 36 in Push-on-Joint Ductile Iron Pipe

Size, in	Pressure class	Thickness, in	Outside diameter,* in	18-ft laying length: Weight per length,† lb	18-ft laying length: Ave. weight lb/ft
3§	350	0.25	3.96	185	9.2
4§	350	0.25	4.80	225	11.3
6§	350	0.25	6.90	300	16.6
8	350	0.25	9.05	395	22.0
10	350	0.26	11.10	510	28.4
12	350	0.28	13.20	655	36.4
14	250	0.28	15.30	770	42.9
	300	0.30	15.30	825	45.8
	350	0.31	15.30	850	47.2
16	250	0.30	17.40	940	52.3
	300	0.32	17.40	1000	55.5
	350	0.34	17.40	1060	58.8
18	250	0.31	19.50	1090	60.5
	300	0.34	19.50	1185	65.9
	350	0.36	19.50	1250	65.9
20	250	0.33	21.60	1290	71.6
	300	0.36	21.60	1395	77.6
	350	0.38	21.60	1470	81.6
24	200	0.33	25.80	1550	86.1
	250	0.37	25.80	1725	95.8
	300	0.40	25.80	1855	103.0
	350	0.43	25.80	1985	110.2
30	150	0.34	32.00	2000	111.2
	200	0.38	32.00	2220	123.2
	250	0.42	32.00	2435	135.2
	300	0.45	32.00	2595	144.2
	350	0.49	32.00	2810	156.1
36	150	0.38	38.30	2675	148.7
	200	0.42	38.30	2935	163.1
	250	0.47	38.30	3260	181.1
	300	0.51	38.30	3520	195.5
	350	0.56	38.30	3840	213.4

*Tolerance of OD of spigot end: 3–12 in, ±0.06 in, 14–24 in, +0.05-in, –0.08 in, 30–36 in, +0.08 in, –0.06 in.

†Including bell; calculated weight of pipe rounded off to nearest 5 lbs

‡Including bell; average weight, per foot, based on calculated weight of pipe before rounding.

§Available in 20-ft lengths.

Equivalent Length of Pipe for 90-Degree Elbows (in Feet)

Velocity, ft/s	Pipe Size														
	1/2	3/4	1	1-1/4	1-1/2	2	2-1/2	3	3-1/2	4	5	6	8	10	12
1	1.2	1.7	2.2	3.0	3.5	4.5	5.4	6.7	7.7	8.6	10.5	12.2	15.4	18.7	22.2
2	1.4	1.9	2.5	3.3	3.9	5.1	6.0	7.5	8.6	9.5	11.7	13.7	17.3	20.8	24.8
3	1.5	2.0	2.7	3.6	4.2	5.4	6.4	8.0	9.2	10.2	12.5	14.6	18.4	22.3	26.5
4	1.5	2.1	2.8	3.7	4.4	5.6	6.7	8.3	9.6	10.6	13.1	15.2	19.2	23.2	27.6
5	1.6	2.2	2.9	3.9	4.5	5.9	7.0	8.7	10.0	11.1	13.6	15.8	19.8	24.2	28.8
6	1.7	2.3	3.0	4.0	4.7	6.0	7.2	8.9	10.3	11.4	14.0	16.3	20.5	24.9	29.6
7	1.7	2.3	3.0	4.1	4.8	6.2	7.4	9.1	10.5	11.7	14.3	16.7	21.0	25.5	30.3
8	1.7	2.4	3.1	4.2	4.9	6.3	7.5	9.3	10.8	11.9	14.6	17.1	21.5	26.1	31.0
9	1.8	2.4	3.2	4.3	5.0	6.4	7.7	9.5	11.0	12.2	14.9	17.4	21.9	26.6	31.6
10	1.8	2.5	3.2	4.3	5.1	6.5	7.8	9.7	11.2	12.4	15.2	17.7	22.2	27.0	32.0

Equivalent Length (Pipe, Elbows, Tees, and Valves)

Find the nominal pipe size being used in the left-most column. For each fitting, read the value under the appropriate heading and add this to the length of piping. This allows total system pressure drop to be calculated. (This is valid for any fluid.)

PIPE SIZE	EQUIVALENT LENGTH OF STRAIGHT PIPE (FEET)				
	STANDARD ELBOW	STANDARD TEE	GATE VALV E FULL OPEN	GLOBE VALVE FULL OPEN	ANGLE VALVE FULL OPEN
1-1/2	4	9	0.9	41	21
2	5	11	1.2	54	27
2-/1/2	6	13	1.4	64	32
3	8	16	1.6	80	40
3-1/2	9	18	2.0	91	45
4	11	21	2.2	110	55
5	13	26	2.8	140	70
6	16	32	3.4	155	81
8	20	42	4.5	210	110
10	25	55	5.5	270	140
12	30	65	6.5	320	160
14	35	75	8.0	370	190

This table contains the number of feet of straight pipe usually allowed for standard fittings and valves.

Metric Equivalent of NPS, ASHRAE, AWWA, NFPA, and ASTM Tube and Pipe Sizes

Nominal pipe size (NPS), in IP	ASHRAE std. wt. size, mm	AWWA pipe size, mm	NFPA pipe size, mm	ASTM copper tube size, mm	Nominal pipe size DN, mm
⅛	—	—	—	6	6
3/16	—	—	—	8	8
¼	8	—	—	10	10
⅜	10	—	—	12	12
½	15	12.7 & 13	12	15	15
⅝	—	—	—	18	18
¾	20	—	—	22	20
1	25	25	25 & 25.4	28	25
1¼	32	—	33	35	32
1½	40	45	38 & 38.1	42	40
2	50	50 & 50.8	51	54	50
2½	65	63 & 63.5	63.5 & 64	67	65
3	80	75	76 & 80	79	80
3½	—	—	89	—	90
4	100	100	102	105	100
4½	—	114.3			115
5	—	—	127	130	125
6	150	150	152	156	150
8	200	200	203	206	200
10	250	250	—	257	250
12	300	300	305	308	300
14	—	350	—		350
18	—	400	—		400
18	—	—	—		450
20	—	500	—		500
24	—	600	—		600
28					700
30					750
32					800
36					900
40					1000
44					1100
48					1200
52					1300
56					1400
60					1500

(*By permission: McGraw-Hill Inc.* Plumber's and Pipefitter's Calculations Manual. *R. Dodge Woodson.*)

One Cubic Yard of Concrete Will Place

Thickness (in)	Square ft	Thickness (in)	Square ft
1	324	7	46
1¼	259	7¼	45
1½	216	7½	43
1¾	185	7¾	42
2	162	8	40
2¼	144	8¼	39
2½	130	8½	38
2¾	118	8¾	37
3	108	9	36
3¼	100	9¼	35
3½	93	9½	34
3¾	86	9¾	33
4	81	10	32.5
4¼	76	10¼	31.5
4½	72	10½	31
4¾	68	10¾	30
5	65	11	29.5
5¼	62	11¼	29
5½	59	11½	28
5¾	56	11¾	27.5
6	54	12	27
6¼	52	15	21.5
6½	50	18	18
6¾	48	24	13.5

Note: By adding just 1 gallon of water to the design mix of 3000 psi concrete, you may:

1. increase slump by 1 in.
2. cut compressive strength by as much as 200 psi
3. increase shrinkage potential by as much as 10%
4. decrease freeze–thaw capability by as much as 20%

Guidelines for Mixing Small Batches of Concrete (by Weight)

Max. size aggregate	Cement (lb/kg)	Wet-fine aggregate (lb/kg)	Wet-coarse aggregate (lb/kg)	Water (lb/kg)
⅜" (9.52 mm)	29 lb (13.15 kg)	59 lb (26.76 kg)	46 lb (20.87 kg)	11 lb (4.99 kg)
½" (12.6 mm)	27 lb (12.25 kg)	53 lb (24.04 kg)	55 lb (24.95 kg)	11 lb (4.99 kg)
¾" (19.05 mm)	25 lb (11.34 kg)	47 lb (21.32 kg)	65 lb (29.66 kg)	10 lb (4.54 kg)
1" (25.39 mm)	24 lb (10.89 kg)	45 lb (20.41 kg)	70 lb (31.75 kg)	10 lb (4.54 kg)
1½" (37.99 mm)	23 lb (10.43 kg)	43 lb (19.50 kg)	75 lb (34.02 kg)	9 lb (4.08 kg)

Guidelines for Mixing Small Batches of Concrete (by Volume)

Max. size aggregate	Cement	Wet-fine aggregate	Wet-coarse aggregate	Water
⅜" (9.52 mm)	1	2½	1½	½
½" (12.6 mm)	1	2½	2	½
¾" (19.05 mm)	1	2½	2½	½
1" (25.39 mm)	1	2½	2¾	½
1½" (37.99 mm)	1	2½	3	½

Recommended Slumps

The Portland Cement Association recommends the following slumps:

Component	Max. slump (inches	Min. slump (inches)
Footings (reinforced or not)	3	1
Foundation walls	3	1
Substructure walls	3	1
Caissons	3	1
Beams and reinforced walls	4	1
Building columns	4	1
Pavements and slabs	3	1
Mass concrete	2	1

BAR SIZE DESIGNATION	WEIGHT POUNDS PER FOOT	NOMINAL DIMENSIONS–ROUND SECTIONS DIAMETER INCHES	CROSS-SECTIONAL AREA-SQ INCHES	PERIMETER INCHES
#3	.376	.375	.11	1.178
#4	.668	.500	.20	1.571
#5	1.043	.625	.31	1.963
#6	1.502	.750	.44	2.356
#7	2.044	.875	.60	2.749
#8	2.670	1.000	.79	3.142
#9	3.400	1.128	1.00	3.544
#10	4.303	1.270	1.27	3.990
#11	5.313	1.410	1.56	4.430
#14	7.650	1.693	2.25	5.320
#18	13.600	2.257	4.00	7.090

Concrete reinforcing bar size/weight chart.

ASTM Standards, Including Soft Metric

Soft metric size	Nom diam mm	Area mm^2	Weight factors kg/m	kg/ft	Imperial size	Nom diam inches	Area in^2	Weight factors lb/ft	lb/m
10	9.5	71	.560	.171	3	.375	.11	.376	1.234
13	12.7	129	.994	.303	4	.500	.20	.668	2.192
16	15.9	199	1.552	.473	5	.625	.31	1.043	3.422
19	19.1	284	2.235	.681	6	.750	.44	1.502	4.928
22	22.2	387	3.042	.927	7	.875	.60	2.044	6.706
25	25.4	510	3.973	1.211	8	1.000	.79	2.670	8.760
29	28.7	645	5.060	1.542	9	1.128	1.00	3.400	11.155
32	32.3	819	6.404	1.952	10	1.270	1.27	4.303	14.117
36	35.8	1006	7.907	2.410	11	1.410	1.56	5.313	17.431
43	43.0	1452	11.384	3.470	14	1.693	2.25	7.650	25.098
57	57.3	2581	20.239	6.169	18	2.257	4.00	13.600	44.619

Comparison of Steel Grades

Soft metric Grade	mPa	psi	Imperial Grade	mPa	psi
300	300	43,511	40	257.79	40,000
420	420	60,716	60	413.69	60,000
520	520	75,420	75	517.11	75,000

(*By permission, Concrete Reinforcing Steel Institute, Schramsburg, Illinois*)

Common Types of Welded Wire Fabric

Style designation (W = Plain, D = Deformed)	Steel area (in ²/ft)		Approximate weight (lb per 100 sq ft)
	Longitudinal	Transverse	
4 x 4-W1.4 x W1.4	0.042	0.042	31
4 x 4-W2.0 x W2.0	0.060	0.060	43
4 x 4-W2.9 x W2.9	0.087	0.087	62
4 x 4-W/D4 x W/D4	0.120	0.120	86
6 x 6-W1.4 x W1.4	0.028	0.028	21
6 x 6-W2.0 x W2.0	0.040	0.040	29
6 x 6-W2.9 x W2.9	0.058	0.058	42
6 x 6-W/D4 x W/D4	0.080	0.080	58
6 x 6-W/D4.7 x W/D4.7	0.094	0.094	68
6 x 6-W/D7.4 x W/D7.4	0.148	0.148	107
6 x 6-W/D7.5 x W/D7.5	0.150	0.150	109
6 x 6-W/D7.8 x W/D7.8	0.156	0.156	113
6 x 6-W/D8 x W/D8	0.160	0.160	116
6 x 6-W/D8.1 x W/D8.1	0.162	0.162	118
6 x 6-W/D8.3 x W/D8.3	0.166	0.166	120
12 x 12-W/D8.3 x W/D8.3	0.083	0.083	63
12 x 12-W/D8.8 x W/D8.8	0.088	0.088	67
12 x 12-W/D9.1 x W/D9.1	0.091	0.091	69
12 x 12-W/D9.4 x W/D9.4	0.094	0.094	71
12 x 12-W/D16 x W/D16	0.160	0.160	121
12 x 12-W/D16.6 x W/D16.6	0.166	0.166	126

*Many styles may be obtained in rolls.

Welded Wire Fabric (WWF)

Wire size number		Nominal diameter, in.	Nominal weight, lb/ft	Area per width (in.^{2}/ft) for various spacings (in)						
Plain	Deformed			2	3	4	6	8	12	16
W45	D45	0.757	1.53	2.70	1.80	1.35	0.90	0.68	0.45	0.34
W31	D31	0.628	1.05	1.86	1.24	0.93	0.62	0.47	0.31	0.23
W20	D20	0.505	0.680	1.2	0.80	0.60	0.40	0.30	0.20	0.15
W18	D18	0.479	0.612	1.1	0.72	0.54	0.36	0.27	0.18	0.14
W16	D16	0.451	0.544	0.96	0.64	0.48	0.32	0.24	0.16	0.12
W14	D14	0.422	0.476	0.84	0.56	0.42	0.28	0.21	0.14	0.11
W12	D12	0.391	0.408	0.72	0.48	0.36	0.24	0.18	0.12	0.09
W11	D11	0.374	0.374	0.66	0.44	0.33	0.22	0.17	0.11	0.08
W10.5		0.366	0.357	0.63	0.42	0.32	0.21	0.16	0.11	0.08
W10	D10	0.357	0.340	0.60	0.40	0.30	0.20	0.15	0.10	0.08
W9.5		0.348	0.323	0.57	0.38	0.29	0.19	0.14	0.095	0.07
W9	D9	0.338	0.306	0.54	0.36	0.27	0.18	0.14	0.090	0.07
W8.5		0.329	0.289	0.51	0.34	0.26	0.17	0.13	0.085	0.06
W8	D8	0.319	0.272	0.48	0.32	0.24	0.16	0.12	0.080	0.06
W7.5		0.309	0.255	0.45	0.30	0.23	0.15	0.11	0.075	0.06
W7	D7	0.299	0.238	0.42	0.28	0.21	0.14	0.11	0.070	0.05
W6.5		0.288	0.221	0.39	0.26	0.20	0.13	0.097	0.065	0.05
W6	D6	0.276	0.204	0.36	0.24	0.18	0.12	0.090	0.060	0.05
W5.5		0.265	0.187	0.33	0.22	0.17	0.11	0.082	0.055	0.04
W5	D5	0.252	0.170	0.30	0.20	0.15	0.10	0.075	0.050	0.04
W4.5		0.239	0.153	0.27	0.18	0.14	0.090	0.067	0.045	0.03
W4	D4	0.226	0.136	0.24	0.16	0.12	0.080	0.060	0.040	0.03
W3.5		0.211	0.119	0.21	0.14	0.11	0.070	0.052	0.035	0.03
W3		0.195	0.102	0.18	0.12	0.090	0.060	0.045	0.030	0.02
W2.9		0.192	0.099	0.17	0.12	0.087	0.058	0.043	0.029	0.02
W2.5		0.178	0.085	0.15	0.10	0.075	0.050	0.037	0.025	0.02
W2.1		0.162	0.070	0.13	0.84	0.063	0.042	0.031	0.021	0.02
W2		0.160	0.068	0.12	0.080	0.060	0.040	0.030	0.020	0.02
W1.5		0.138	0.051	0.090	0.060	0.045	0.030	0.022	0.015	0.01
W1.4		0.134	0.048	0.084	0.056	0.042	0.028	0.021	0.014	0.01

(By permission, Concrete Reinforcing Steel Institute, Schramsburg, Illinois)

Mortar Types

- *Type M* High compressive strength (2500 psi average), containing greater durability than other types. Therefore, it is generally recommended for unreinforced masonry walls below grade.
- *Type S* Reasonable high compressive strength (1800 psi average) and having great tensile bond strength. It is usually recommended for reinforced masonry walls, where maximum flexural strength is required.
- *Type N* Midrange compressive strength (750 psi average) and suitable for general above-grade masonry construction for parapets and chimneys.
- *Type O* Load compressive strength (350 psi average) and suitable for interior non-load-bearing masonry walls.
- *Type K* Very low compressive strength (75 psi average) and occasionally used for interior non-load-bearing walls, where permitted by local building codes.

Workability or plasticity of the mortar is an essential characteristic of proper mortar mixes. The mortar must have both cohesive and adhesive qualities when it makes contact with the masonry units. Hardness or high strength is not necessarily a measure of durability. Mortar that is stronger than the masonry units to which it is applied might not "give," thereby causing stress to be relieved by the masonry units. This could result in these units cracking or spalling.

Compressive Strength of Mortars Made with Various Types of Cement

Type of cement	Minimum compressive strength, psi				ASTM designation
	1 day	3 days	7 days	28 days	
Portland cements					C150-85
I	—	1800	2800	4000*	
IA	—	1450	2250	3200*	
II	—	1500	2500	4000*	
	—	1000†	1700†	3200*†	
IIA	—	1200	2000	3200*	
	—	800†	1350†	2560*†	
III	1800	3500	—	—	
IIIA	1450	2800	—	—	
IV	—	—	1000	2500	
V	—	1200	2200	3000	
Blended cements					C595-85
I(SM), IS, I(PM), IP	—	1800	2800	3500	
I(SM)-A, IS-A I(PM)-A, IP-A	—	1450	2250	2800	
IS(MS), IP(MS)	—	1500	2500	3500	
IS-A(MS), IP-A(MS)	—	1200	2000	2800	
S	—	—	600	1500	
SA	—	—	500	1250	
P	—	—	1500	3000	
PA	—	—	1250	2500	
Expansive cement					C845-80
E-1	—	—	2100	3500	
Masonry cements					C91-83a
N	—	—	500	900	
S	—	—	1300	2100	
M	—	—	1800	2900	

*Optional requirement.

†Applicable when the optional heat of hydration or chemical limit on the sum of C2S and C3A is specified.

Note: When low or moderate heat of hydration is specified for blended cements (ASTM C595), the strength requirements is 80% of the value shown.

(By permission from the Masonry Society, ACI, ASCE from their manual Building Code Requirements for Masonry Structures.)

Proportions by Volume of M, S, N, and O Mortars

MORTAR	TYPE	PROPORTIONS BY VOLUME (CEMENTITIOUS MATERIALS)								AGGREGATE MEASURED IN A DAMP, LOOSE CONDITION
		Portland Cement or Blended Cement	Masonry Cement[1]			Mortar Cement[2]			Hydrated Lime or Lime Putty	
			M	S	N	M	S	N		
Cement-lime	M	1	—	—	—	—	—	—	1/4	
	S	1	—	—	—	—	—	—	over 1/4 to 1/2	
	N	1	—	—	—	—	—	—	over 1/2 to 1 1/4	
	O	1	—	—	—	—	—	—	over 1 1/4 to 2 1/2	
Mortar cement	M	1	—	—	—	—	—	1	—	Not less than 2 1/4 and not more than 3 times the sum of the separate volumes of cementitious materials.
	M	—	—	—	—	1	—	—	—	
	S	1/2	—	—	—	—	—	1	—	
	S	—	—	—	—	—	1	—	—	
	N	—	—	—	—	—	—	1	—	
Masonry cement	M	1	—	—	1	—	—	—	—	
	M	—	1	—	—	—	—	—	—	
	S	1/2	—	—	1	—	—	—	—	
	S	—	—	1	—	—	—	—	—	
	N	—	—	—	1	—	—	—	—	
	O	—	—	—	1	—	—	—	—	

[1]Masonry cement conforming to the requirements of UBC Standard 21-11.

[2]Mortar cement conforming to the requirements of UBC Standard 21-14.

Estimating Concrete Masonry

NOMINAL LENGTH OF CONCRETE MASONRY WALLS BY STRETCHERS

(Based on units 15⅝" long and half units 7⅝" long with ⅜" thick head joints)

LENGTH OF WALL	NO. OF UNITS	LENGTH OF WALL	NO. OF UNITS	LENGTH OF WALL	NO. OF UNITS	LENGTH OF WALL	NO. OF UNITS	LENGTH OF WALL	NO. OF UNITS	LENGTH OF WALL	NO. OF UNITS
0′-8″	½	20′-8″	15½	40′-8″	30½	60′-8″	45½	80′-8″	60½	100′-8″	75½
1′-4″	1	21′-4″	16	41′-4″	31	61′-4″	46	81′-4″	61	101′-4″	76
2′-0″	1½	22′-0″	16½	42′-0″	31½	62′-0″	46½	82′-0″	61½	102′-0″	76½
2′-8″	2	22′-8″	17	42′-8″	32	62′-8″	47	82′-8″	62	102′-8″	77
3′-4″	2½	23′-4″	17½	43′-4″	32½	63′-4″	47½	83′-4″	62½	103′-4″	77½
4′-0″	3	24′-0″	18	44′-0″	33	64′-0″	48	84′-0″	63	104′-0″	78
4′-8″	3½	24′-8″	18½	44′-8″	33½	64′-8″	48½	84′-8″	63½	104′-8″	78½
5′-4″	4	25′-4″	19	45′-4″	34	65′-4″	49	85′-4″	64	105′-4″	79
6′-0″	4½	26′-0″	19½	46′-0″	34½	66′-0″	49½	86′-0″	64½	106′-0″	79½
6′-8″	5	26′-8″	20	46′-8″	35	66′-8″	50	86′-8″	65	106′-8″	80
7′-4″	5½	27′-4″	20½	47′-4″	35½	67′-4″	50½	87′-4″	65½	107′-4″	80½
8′-0″	6	28′-0″	21	48′-0″	36	68′-0″	51	88′-0″	66	108′-0″	81
8′-8″	6½	28′-8″	21½	48′-8″	36½	68′-8″	51½	88′-8″	66½	108′-8″	81½
9′-4″	7	29′-4″	22	49′-4″	37	69′-4″	52	89′-4″	67	109′-4″	82
10′-0″	7½	30′-0″	22½	50′-0″	37½	70′-0″	52½	90′-0″	67½	110′-0″	82½
10′-8″	8	30′-8″	23	50′-8″	38	70′-8″	53	90′-8″	68	110′-8″	83
11′-4″	8½	31′-4″	23½	51′-4″	38½	71′-4″	53½	91′-4″	68½	111′-4″	83½
12′-0″	9	32′-0″	24	52′-0″	39	72′-0″	54	92′-0″	69	112′-0″	84
12′-8″	9½	32′-8″	24½	52′-8″	39½	72′-8″	54½	92′-8″	69½	112′-8″	84½
13′-4″	10	33′-4″	25	53′-4″	40	73′-4″	55	93′-4″	70	113′-4″	85
14′-0″	10½	34′-0″	25½	54′-0″	40½	74′-0″	55½	94′-0″	70½	114′-0″	85½
14′-8″	11	34′-8″	26	54′-8″	41	74′-8″	56	94′-8″	71	114′-8″	86
15′-4″	11½	35′-4″	26½	55′-4″	41½	75′-4″	56½	95′-4″	71½	115′-4″	86½
16′-0″	12	36′-0″	27	56′-0″	42	76′-0″	57	96′-0″	72	116′-0″	87
16′-8″	12½	36′-8″	27½	56′-8″	42½	76′-8″	57½	96′-8″	72½	116′-8″	87½
17′-4″	13	37′-4″	28	57′-4″	43	77′-4″	58	97′-4″	73	117′-4″	88
18′-0″	13½	38′-0″	28½	58′-0″	43½	78′-0″	58½	98′-0″	73½	118′-0″	88½
18′-8″	14	38′-8″	29	58′-8″	44	78′-8″	59	98′-8″	74	118′-8″	89
19′-4″	14½	39′-4″	29½	59′-4″	44½	79′-4″	59½	99′-4″	74½	119′-4″	89½
20′-0″	15	40′-0″	30	60′-0″	45	80′-0″	60	100′-0″	75	120′-0″	90

NOMINAL HEIGHT OF CONCRETE MASONRY WALLS BY COURSES

(Based on units 7⅝" high ⅜" thick mortar joints)

HEIGHT OF WALL	NO. OF UNITS	HEIGHT OF WALL	NO. OF UNITS	HEIGHT OF WALL	NO. OF UNITS	HEIGHT OF WALL	NO. OF UNITS
0′-8″	1	8′-8″	13	16′-8″	25	24′-8″	37
1′-4″	2	9′-4″	14	17′-4″	26	25′-4″	38
2′-0″	3	10′-0″	15	18′-0″	27	26′-0″	39
2′-8″	4	10′-8″	16	18′-8″	28	26′-8″	40
3′-4″	5	11′-4″	17	19′-4″	29	27′-4″	41
4′-0″	6	12′-0″	18	20′-0″	30	28′-0″	42
4′-8″	7	12′-8″	19	20′-8″	31	28′-8″	43
5′-4″	8	13′-4″	20	21′-4″	32	29′-4″	44
6′-0″	9	14′-0″	21	22′-0″	33	30′-0″	45
6′-8″	10	14′-8″	22	22′-8″	34	30′-8″	46
7′-4″	11	15′-4″	23	23′-4″	35	31′-4″	47
8′-0″	12	16′-0″	24	24′-0″	36	30′-0″	48

HOW TO USE THESE TABLES

The tables on this page are an aid to estimating and designing with standard concrete masonry units. The following are examples of how they can be used to advantage.

Example:

Estimate the number of units required for a wall 76′ long and 12′ high.

From table: 76′ = 57 units
12′ = 18 courses

57 × 18 = 1026 = No. masonry units required

Example:

Estimate the number of units required for a foundation 24′ × 30′ = 11 courses high.

2 (24 + 30) = 108′ = distance for a foundation

From table: 108′ = 81 units

81 × 11 = 891 = No. masonry units required.

This table can also be useful in the layout of a building on a modular basis to eliminate cutting of units. Example: If design calls for a wall 41′ long it can be found from the table that making wall 41′-4″, will eliminate cutting units and consequent waste. Example: If the distance between two openings has been tentatively established at 2′-9″, consulting the table will show that 2′-8″ dimension would eliminate cutting of units.

Nominal Height of Brick and Block Walls by Coursing

COURSES	REGULAR 4 2 1/4" bricks + 4 equal joints =					MODULAR 3 bricks + 3 joints =	CONCRETE BLOCKS	
	10" 1/4" joints	10 1/2" 3/8" joints	11" 1/2" joints	11 1/2" 5/8" joints	12" 3/4" joints	8"	3 5/8" blocks 3/8" joints	7 5/8" blocks 3/8" joints
1	2 1/2"	2 5/8"	2 3/4"	2 7/8"	3"	2 11/16"	4"	8"
2	5"	5 1/4"	5 1/2"	5 3/4"	6"	5 5/16"	8"	1'4"
3	7 1/2"	7 7/8"	8 1/4"	8 5/8"	9"	8"	1'0"	2'0"
4	10"	10 1/2"	11"	11 1/2"	1'0"	10 11/16"	1'4"	2'8"
5	1'0 1/2"	1'1 1/8"	1'1 3/4"	1'2 3/8"	1'3"	1'1 5/16"	1'8"	3'4"
6	1'3"	1'3 3/4"	1'4 1/2"	1'5 1/4"	1'6"	1'4"	2'0"	4'0"
7	1'5 1/2"	1'6 3/8"	1'7 1/4"	1'8 1/8"	1'9"	1'6 11/16"	2'4"	4'8"
8	1'8"	1'9"	1'10"	1'11"	2'0"	1'9 5/16"	2'8"	5'4"
9	1'10 1/2"	1'11 5/8"	2'0 3/4"	2'1 7/8"	2'3"	2'0"	3'0"	6'0"
10	2'1"	2'2 1/4"	2'3 1/2"	2'4 3/4"	2'6"	2'2 11/16"	3'4"	6'8"
11	2'3 1/2"	2'4 7/8"	2'6 1/4"	2'7 5/8"	2'9"	2'5 5/16"	3'8"	7'4"
12	2'6"	2'7 1/2"	2'9"	2'10 1/2"	3'0"	2'8"	4'0"	8'0"
13	2'8 1/2"	2'10 1/8"	2'11 3/4"	3'1 3/8"	3'3"	2'10 11/16"	4'4"	8'8"
14	2'11"	3'0 3/4"	3'2 1/2"	3'4 1/4"	3'6"	3'1 5/16"	4'8"	9'4"
15	3'1 1/2"	3'3 3/8"	3'5 1/4"	3'7 1/8"	3'9"	3'4"	5'0"	10'0"
16	3'4"	3'6"	3'8"	3'10"	4'0"	3'6 11/16"	5'4"	10'8"
17	3'6 1/2"	3'8 5/8"	3'10 3/4"	4'0 7/8"	4'3"	3'9 5/16"	5'8"	11'4"
18	3'9"	3'11 1/4"	4'1 1/2"	4'3 3/4"	4'6"	4'0"	6'0"	12'0"
19	3'11 1/2"	4'1 7/8"	4'4 1/4"	4'6 5/8"	4'9"	4'2 11/16"	6'4"	12'8"
20	4'2"	4'4 1/2"	4'7"	4'9 1/2"	5'0"	4'5 5/16"	6'8"	13'4"
21	4'4 1/2"	4'7 1/8"	4'9 3/4"	5'0 3/8"	5'3"	4'8"	7'0"	14'0"
22	4'7"	4'9 3/4"	5'0 1/2"	5'3 1/4"	5'6"	4'10 11/16"	7'4"	14'8"
23	4'9 1/2"	5'0 3/8"	5'3 1/4"	5'6 1/8"	5'9"	5'1 5/16"	7'8"	15'4"
24	5'0"	5'3"	5'6"	5'9"	6'0"	5'4"	8'0"	16'0"
25	5'2 1/2"	5'5 5/8"	5'8 3/4"	5'11 7/8"	6'3"	5'6 11/16"	8'4"	16'8"
26	5'5"	5'8 1/4"	5'11 1/2"	6'2 3/4"	6'6"	5'9 5/16"	8'8"	17'4"
27	5'7 1/2"	5'10 7/8"	6'2 1/4"	6'5 5/8"	6'9"	6'0"	9'0"	18'0"
28	5'10"	6'1 1/2"	6'5"	6'8 1/2"	7'0"	6'2 11/16"	9'4"	18'8"
29	6'0 1/2"	6'4 1/8"	6'7 3/4"	6'11 3/8"	7'3"	6'5 5/16"	9'8"	19'4"
30	6'3"	6'6 3/4"	6'10 1/2"	7'2 1/4"	7'6"	6'8"	10'0"	20'0"
31	6'5 1/2"	6'9 3/8"	7'1 1/4"	7'5 1/8"	7'9"	6'10 11/16"	10'4"	20'8"
32	6'8"	7'0"	7'4"	7'8"	8'0"	7'1 5/16"	10'8"	21'4"
33	6'10 1/2"	7'2 5/8"	7'6 3/4"	7'10 7/8"	8'3"	7'4"	11'0"	22'0"
34	7'1"	7'5 1/4"	7'9 1/2"	8'1 3/4"	8'6"	7'6 11/16"	11'4"	22'8"
35	7'3 1/2"	7'7 7/8"	8'0 1/4"	8'4 5/8"	8'9"	7'9 5/16"	11'8"	23'4"
36	7'6"	7'10 1/2"	8'3"	8'7 1/2"	9'0"	8'0"	12'0"	24'0"
37	7'8 1/2"	8'1 1/8"	8'5 3/4"	8'10 3/8"	9'3"	8'2 11/16"	12'4"	24'8"
38	7'11"	8'3 3/4"	8'8 1/2"	9'1 1/4"	9'6"	8'5 5/16"	12'8"	25'4"
39	8'1 1/2"	8'6 3/8"	8'11 1/4"	9'4 1/8"	9'9"	8'8"	13'0"	26'0"
40	8'4"	8'9"	9'2"	9'7"	10'0"	8'10 11/16"	13'4"	26'8"
41	8'6 1/2"	8'11 5/8"	9'4 3/4"	9'9 7/8"	10'3"	9'1 5/16"	13'8"	27'4"
42	8'9"	9'2 1/4"	9'7 1/2"	10'0 3/4"	10'6"	9'4"	14'0"	28'0"
43	8'11 1/2"	9'4 7/8"	9'10 1/4"	10'3 5/8"	10'9"	9'6 11/16"	14'4"	28'8"
44	9'2"	9'7 1/2"	10'1"	10'6 1/2"	11'0"	9'9 5/16"	14'8"	29'4"
45	9'4 1/2"	9'10 1/8"	10'3 3/4"	10'9 3/8"	11'3"	10'0"	15'0"	30'0"
46	9'7"	10'0 3/4"	10'6 1/2"	11'0 1/4"	11'6"	10'2 11/16"	15'4"	30'8"
47	9'9 1/2"	10'3 3/8"	10'9 1/4"	11'3 1/8"	11'9"	10'5 5/16"	15'8"	31'4"
48	10'0"	10'6"	11'0"	11'6"	12'0"	10'8"	16'0"	32'0"
49	10'2 1/2"	10'8 5/8"	11'2 3/4"	11'8 7/8"	12'3"	10'10 11/16"	16'4"	32'8"
50	10'5"	10'11 1/4"	11'5 1/2"	11'11 3/4"	12'6"	11'1 5/16"	16'8"	33'4"

History of Steel and Grades of Structural Steel

Iron was produced by primitive man by placing iron ore and charcoal in a clay pot and building a fire in the pot, using a crude bellows to provide the forced draft that deposited iron at the bottom. It was not until the mid-1800s that Henry Bessemer, an English metallurgist, developed a process whereby forced air was introduced into the iron-refining procedure raising the temperature of the crucible so that impurities in the molten pig iron were burned away. In the process, a more malleable metal, steel, was created.

Various minerals and metals are added to molten steel nowadays to enhance certain characteristics:

- *Nickel* Improves the hardenability of steel and increases impact strength at low temperatures.
- *Sulfur* Increases machinability.
- *Manganese* Increases strength and hardness.
- *Carbon* The principal hardening agent in steel.
- *Molydenum* Prevents brittleness.
- *Vanadium* Gives steel a fine grain structure and improves the fatigue values.
- *Silicon* Improves strength. It is a deoxidizer.
- *Phosphorous* Improves the machinability of high-sulfur steel sand imparts some resistance to corrosion.

ASTM Structural Steel Specifications

ASTM designation	Steel type
A36	Carbon
A529	Carbon
A441	High strength (low alloy)
A572 grade (includes 42, 50, 60, 65)	High strength (low alloy)
A242	Corrosion resistant, high strength Low alloy
A588	Corrosion resistant, high strength Low alloy
A852	Quenched and tempered (low alloy) (Plates only)
A514	Quenched and tempered alloy (Plates only)

Welded Joints—Standard Symbols

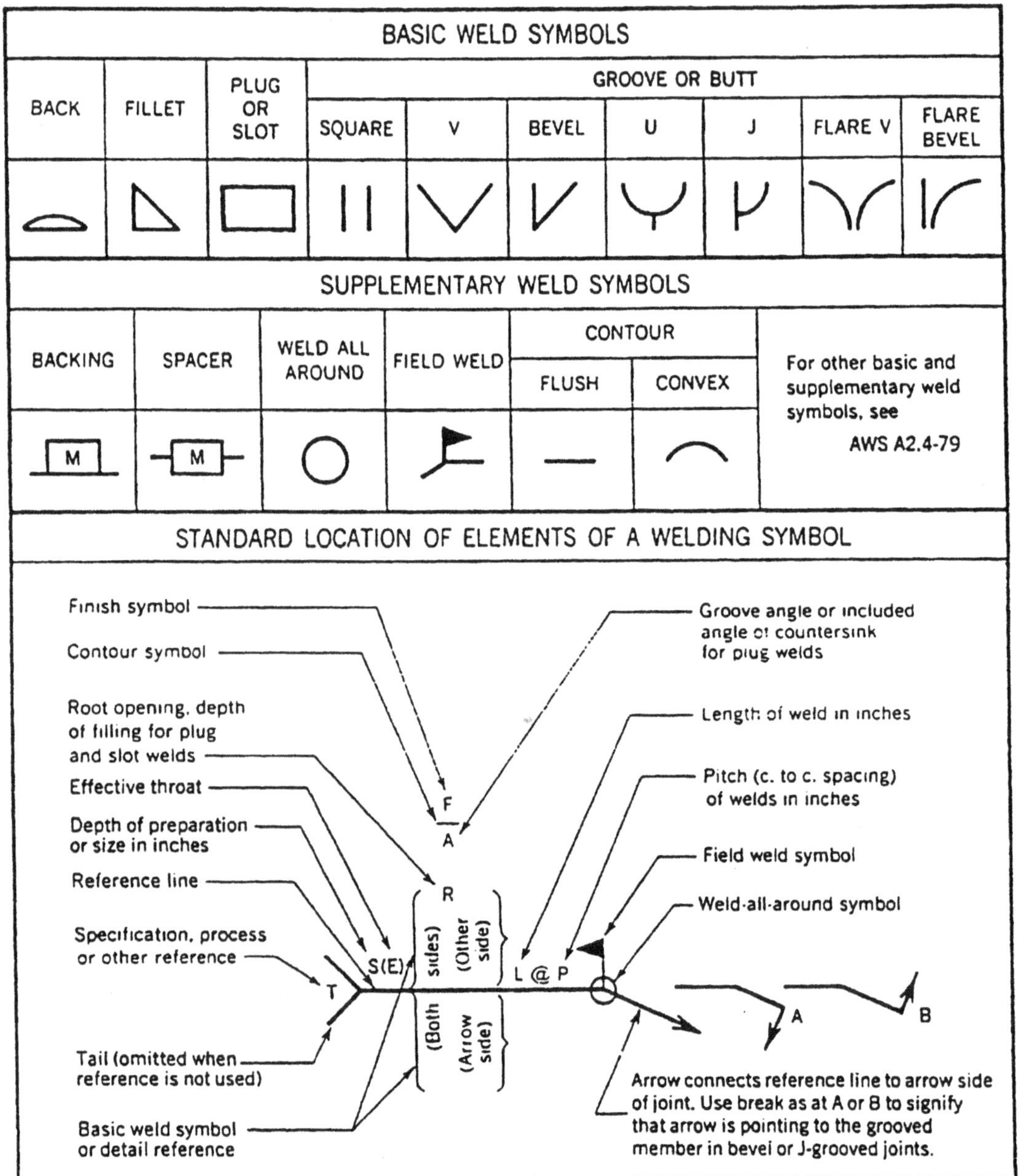

Note:

Size, weld symbol, length of weld and spacing must read in that order from left to right along the reference line. Neither orientation of reference line nor location of the arrow alter this rule.

The perpendicular leg of 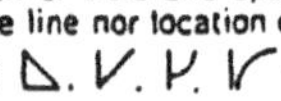 weld symbols must be at left.

Arrow and Other Side welds are of the same size unless otherwise shown. Dimensions of fillet welds must be shown on both the Arrow Side and the Other Side Symbol.

The point of the field weld symbol must point toward the tail.

Symbols apply between abrupt changes in direction of welding unless governed by the "all around" symbol or otherwise dimensioned.

These symbols do not explicitly provide for the case that frequently occurs in structural work, where duplicate material (such as stiffeners) occurs on the far side of a web or gusset plate. The fabricating industry has adopted this convention: that when the billing of the detail material discloses the existence of a member on the far side as well as on the near side, the welding shown for the near side shall be duplicated on the far side.

(By permission of the American Institute of Steel Construction, Chicago, Illinois.)

WWPA Guide to Understanding Grade Stamps

Interpreting Grade Marks

Most grade stamps, except those for rough lumber or heavy timbers, contain 5 basic elements:

a. **WWPA certification mark.** Certifies Association quality supervision. WWP® is a registered trademark.

b. **Mill identification.** Firm name, brand or assigned mill number. WWPA can be contacted to identify an individual mill whenever necessary.

c. **Grade designation.** Grade name, number or abbreviation.

d. **Species identification.** Indicates species by individual species or species combination. Species identification marks for groups to which design values are assigned are:

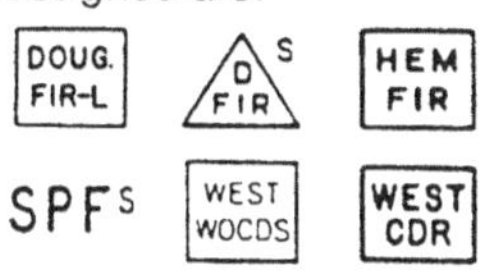

e. **Condition of seasoning.** Indicates condition of seasoning at time of surfacing:

MC-15 — KD-15	15% maximum moisture content
S-DRY — KD	19% maximum moisture content
S-GRN —	over 19% moisture content (unseasoned)

Inspection Certificate

When an inspection certificate issued by the Western Wood Products Association is required on a shipment of lumber and specific grade marks are not used, the stock is identified by an imprint of the Association mark and the number of the shipping mill or inspector.

Grade Stamp Facsimiles

WWPA uses a set of marks similar to the randomly selected examples shown on the reverse side, to identify lumber graded under its supervision.

Species Combinations

The species groupings for dimension lumber products are shown left and explained in the second box on the reverse side. When alternative species combinations, as shown in the third box on the reverse side, are used for structural applications, design values are controlled by the species with the lowest strength value within the combination.

(By permission of Western Wood Products Association.)

Standard Sizes for Western Wood Finish and Selects (Dry Lumber)

The metric dimensions listed in these rules are calculated at 25.4 millimeters (mm) times the actual dimension in inches, rounded to the nearest millimeter. In case of a dispute on size measurements, the conventional (inch) method of measurement shall take precedence.

STANDARD SIZES for FINISH DRY LUMBER

Thicknesses			Widths		
Nominal	Surfaced		Nominal	Surfaced	
	Inch	mm [1]		Inch	mm [1]
3/8"	5/16	8	2"	1 1/2	38
1/2"	7/16	11	3"	2 1/2	64
5/8"	9/16	14	4"	3 1/2	89
3/4"	5/8	16	5"	4 1/2	114
1"	3/4	19	6"	5 1/2	140
1 1/4"	1	25	7"	6 1/2	165
1 1/2"	1 1/4	32	8" and wider	3/4 off nominal	19 off nominal
1 3/4"	1 3/8	35			
2"	1 1/2	38			
2 1/2"	2	51			
3"	2 1/2	64			
3 1/2"	3	76			
4"	3 1/2	89			

STANDARD SIZES for SELECTS DRY LUMBER

Thicknesses			Widths		
Nominal	Surfaced		Nominal	Surfaced	
	Inch	mm [1]		Inch	mm [1]
4/4	3/4	19	2"	1 1/2	38
5/4	1 5/32	29	3"	2 1/2	64
6/4	1 13/32	36	4"	3 1/2	89
7/4	1 19/32	40	5"	4 1/2	114
8/4	1 13/16	46	6"	5 1/2	140
9/4	2 3/32	53	7"	6 1/2	165
10/4	2 3/8	60	8" and wider	3/4 off nominal	19 off nominal
11/4	2 9/16	65			
12/4	2 3/4	70			
16/4	3 3/4	95			

(By permission of Western Wood Products Association.)

Standard Sizes for Western Wood Common Boards, Studs, and Battens

STANDARD SIZES for COMMON BOARDS (Including Thick Lumber Shipped Under Board Rules) DRY LUMBER

Thicknesses			Widths		
Nominal	Surfaced		Nominal	Surfaced	
	Inch	mm (1)		Inch	mm (1)
3/4	$\frac{5}{8}$	16	2"	$1\frac{1}{2}$	38
4/4	$\frac{3}{4}$	19	3"	$2\frac{1}{2}$	64
5/4	$1\frac{5}{32}$	29	4"	$3\frac{1}{2}$	89
6/4	$1\frac{13}{32}$	36	5"	$4\frac{1}{2}$	114
7/4	$1\frac{19}{32}$	40	6"	$5\frac{1}{2}$	140
8/4	$1\frac{13}{16}$	46	7"	$6\frac{1}{2}$	165
9/4	$2\frac{3}{32}$	53	8" and wider	$\frac{3}{4}$ off nominal	19 off nominal
10/4	$2\frac{3}{8}$	60			
11/4	$2\frac{9}{16}$	65			
12/4	$2\frac{3}{4}$	70			
16/4	$3\frac{3}{4}$	95			

Surfaced square size shall be governed by thickness.
At manufacturer's option, dry 4/4 may be $\frac{25}{32}$"
Standard lengths are 6' and longer in multiples of 1.

STANDARD SIZES for STUDS

Nominal	Surfaced Dry		Surfaced Unseasoned	
	Inch	mm (1)	Inch	mm (1)
Thicknesses				
2"	$1\frac{1}{2}$	38	$1\frac{9}{16}$	40
3"	$2\frac{1}{2}$	64	$2\frac{9}{16}$	65
4"	$3\frac{1}{2}$	89	$3\frac{9}{16}$	90
Widths				
2"	$1\frac{1}{2}$	38	$1\frac{9}{16}$	40
3"	$2\frac{1}{2}$	64	$2\frac{9}{16}$	65
4"	$3\frac{1}{2}$	89	$3\frac{9}{16}$	90
5"	$4\frac{1}{2}$	114	$4\frac{5}{8}$	117
6"	$5\frac{1}{2}$	140	$5\frac{5}{8}$	143
8" and wider	$\frac{3}{4}$ off nominal	19 off nominal	$\frac{1}{2}$ off nominal	13 off nominal

BATTENS

All Species

Standard widths are:

Nominal	Net	
	Inch	mm (1)
Flat Battens—3"	$\frac{1}{4} \times 2\frac{1}{2}$	6 x 64
O.G. Battens—2"	$\frac{3}{4} \times 1\frac{3}{4}$	19 x 44
O.G. Battens—$2\frac{1}{2}$"	$\frac{3}{4} \times 2\frac{1}{4}$	19 x 57
O.G. Battens—3"	$\frac{3}{4} \times 2\frac{1}{2}$	19 x 64

(By permission of Western Wood Products Association.)

Knots and How to Measure Them

Measuring of knots for Southern Pine lumber.

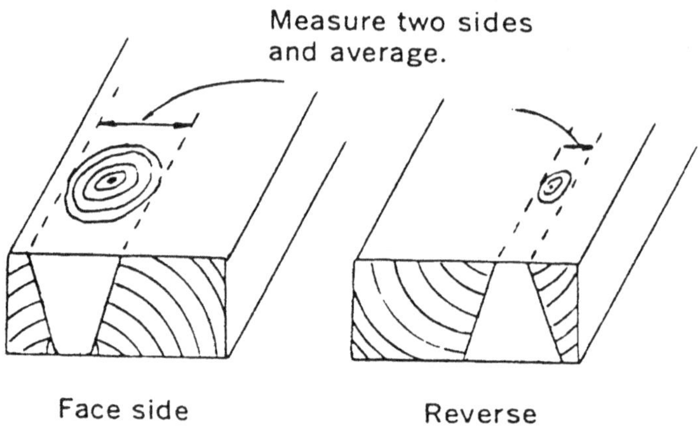

GRADE	KNOT SIZE ON NARROW FACE	KNOT MAY EXTEND INTO PIECE
No. 1	1-1/2"	1/2 of width (Fig. 1)
	1"	3/4 of width (Fig. 2)
No. 2	1-1/2"	Slightly less than 3/4 of width.
No. 3	1-1/2"	Slightly less than width

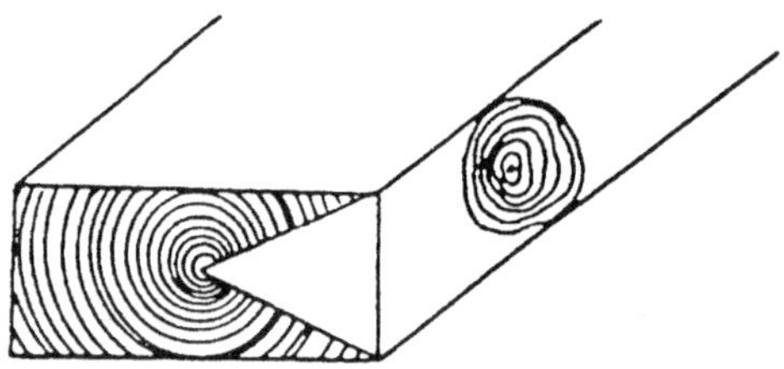

1-1/2" Narrow face knot extending 1/2 of width.

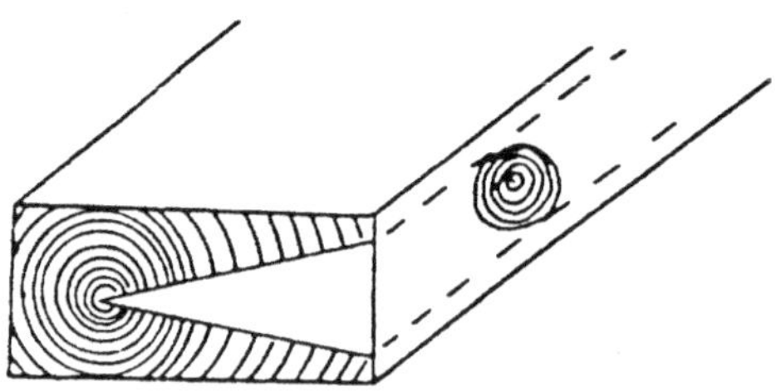

1" Narrow face knot extending 3/4 of width.

(By permission of the Southern Pine Council.)

6.28.0 Plywood Panel Dimensions (U.S. Customary and Metric)

Metric Conversions

Metric equivalents of nominal thicknesses and common sizes of APA Rated Siding products are tabulated below. (1 inch = 25.4 millimeters):

APA RATED SIDING
NOMINAL THICKNESS

in.	mm
11/32	8.7
3/8	9.5
7/16	11.1
15/32	11.9
1/2	12.7
19/32	15.1
5/8	15.9

PANEL SIDING
NOMINAL DIMENSIONS
(Width x Length)

ft	mm	m (approx.)
4 x 8	1219 x 2438	1.22 x 2.44
4 x 9	1219 x 2743	1.22 x 2.74
4 x 10	1219 x 3048	1.22 x 3.05

LAP SIDING
NOMINAL DIMENSIONS
(Width x Length)

in. x ft	mm	m (approx.)
6 x 16	152.4 x 4877	0.15 x 4.88
8 x 16	203.2 x 4877	0.20 x 4.88
12 x 16	304.8 x 4877	0.30 x 4.88

(By permission of APA, The Engineered Wood Association, Tacoma, Washington.)

I-Joist Series—Size, Depth, Flange Width

DIMENSIONS FOR APA PERFORMANCE RATED I-JOISTS

APA PRI				
Joist Series	Joist Designation	Nominal Depth	Net Depth	Flange Width
I x 10	C4 (PRI-15)	10"	9-1/2"	1-1/2"
I x 10	C6 (PRI-25)	10"	9-1/2"	1-3/4"
I x 12	C10 (PRI-15)	12"	11-7/8"	1-1/2"
I x 12	C12 (PRI-25)	12"	11-7/8"	1-3/4"
I x 14	C14 (PRI-25)	14"	14"	1-3/4"
I x 14	C16 (PRI-35)	14"	14"	2-5/16"
I x 16	C18 (PRI-25)	16"	16"	1-3/4"
I x 16	C20 (PRI-35)	16"	16"	2-5/16"
I x 10	S2 (PRI-30)	10"	9-1/2"	2-1/2"
I x 10	S4 (PRI-32)	10"	9-1/2"	2-1/2"
I x 12	S6 (PRI-30)	12"	11-7/8"	2-1/2"
I x 12	S8 (PRI-32)	12"	11-7/8"	2-1/2"
I x 12	S10 (PRI-42)	12"	11-7/8"	3-1/2"
I x 14	S12 (PRI-32)	14"	14"	2-1/2"
I x 14	S14 (PRI-42)	14"	14"	3-1/2"
I x 16	S16 (PRI-32)	16"	16"	2-1/2"
I x 16	S18 (PRI-42)	16"	16"	3-1/2"

Notes:

1. Tolerances permitted at time of manufacture:
   - flange width = +/– 1/32 inch
   - I-joist depth = + 0 inches, – 1/8 inch

(By permission from APA—The Engineered Wood Association, Tacoma, Washington.)

Truss Types

There are two basic types of trusses. The Pitched-Chord Truss is characterized by its usually triangular shape. Also called a "Roof Truss," this type is most often used for roof construction. The Parallel-Chord Truss gets it name from having parallel top and bottom chords. This type is often used for floor construction, and is referred to as a "Floor Truss." Shown below are 12 of the most common types of trusses used today. As can be seen, there is a great deal of flexibility that can be achieved in truss design. You may even recognize common uses of these trusses based on their shapes. Note: truss plates are not shown for simplicity.

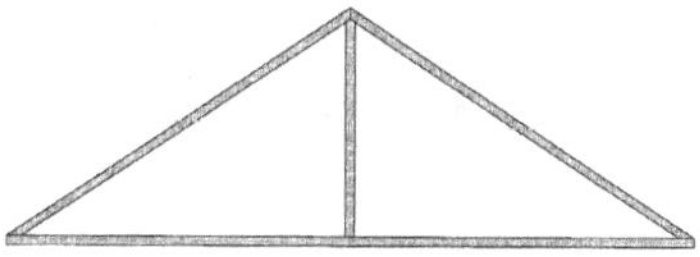

King Post. The most basic truss type. Used primarily for simple structures with short spans.

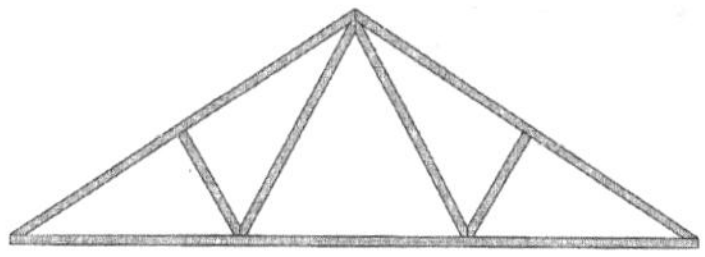

Fink. Provides for longer spans than the King Post, adding further design flexibility.

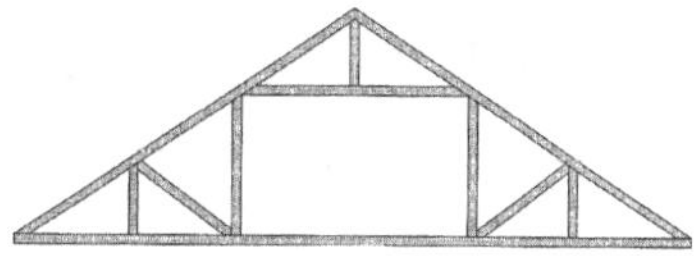

Attic Frame. The opening in the center makes this useful for storage or additional living space applications.

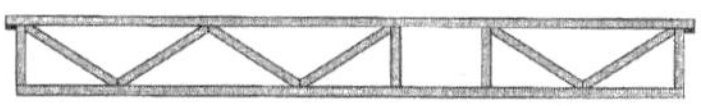

Flat - Top Chord Bearing. The top chord bearing feature can reduce the amount of wall area that needs to be sheathed.

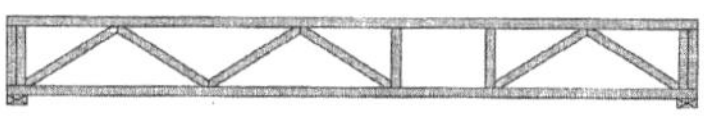

Flat - Bottom Chord Bearing. Also called parallel chord. The 4x2 configuration can be used for floors, and the 2x4 configuration can be used for either floor or roof applications.

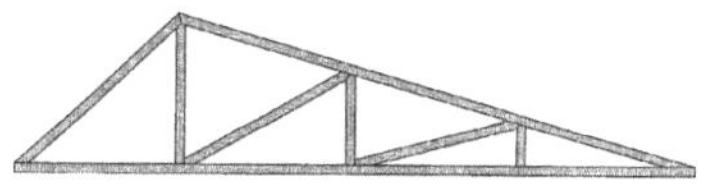

Double Pitch or Dual Pitch. An asymmetrical truss used where the designer wants a change in roof appearance. This truss begins to show the design flexibility roof truss construction provides.

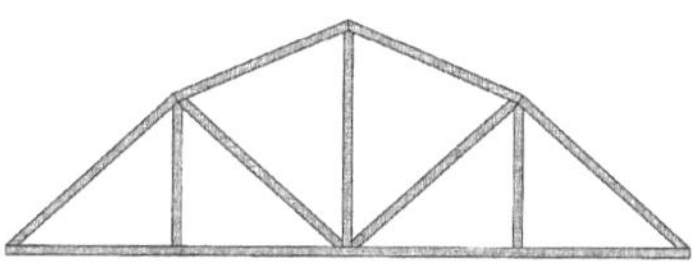

Gambrel. This truss is often used in agricultural buildings.

Hip. One piece of a hip roof system. See page 6 for details.

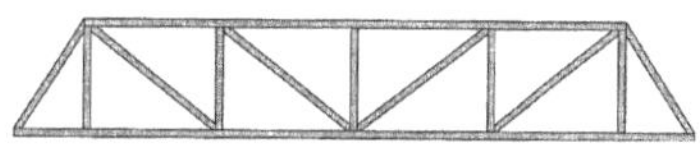

Mansard. This truss is used primarily for commercial buildings where a sloping roof facade hides mechanical units.

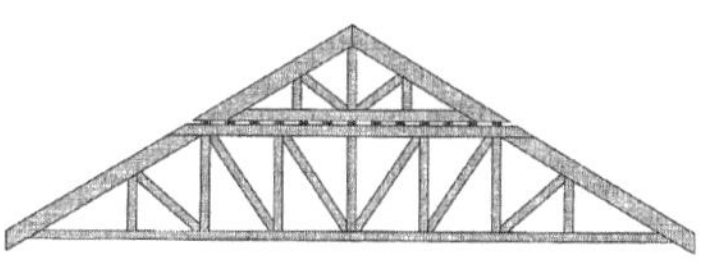

Piggyback. A combination of hip truss and triangular cap, overcoming size restrictions in manufacturing and shipping.

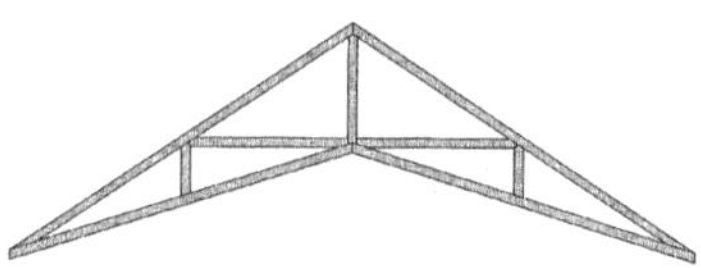

Scissors. Both top and bottom chords of this truss have a pitch to them. Often used to create a vaulted effect, such as in sanctuaries and auditoriums.

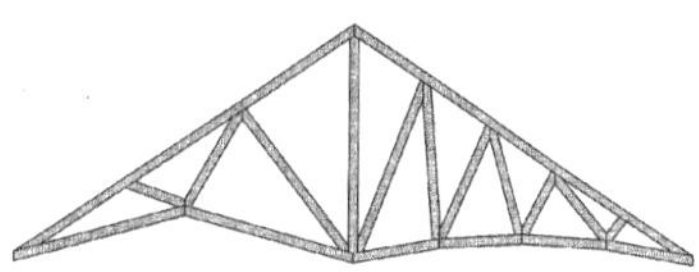

Special. This example illustrates the flexibility of design that trusses will accommodate. The only limit to the design is your imagination!

(Courtesy: Wood Truss Council of America, Madison, Wisconsin.)

A Typical Truss Drawing for a Parallel Chord Truss

F

Job Name: MADISON | **Truss ID: FT1** | **Qty: 18** | **Drwg:**

BRG	X-LOC	REACT	SIZE	REQD
1	0- 1-12	1100	3.50"	1.50"
2	19-10- 4	1100	3.50"	1.50"

TOP CHORD 4x2 SPF 2400F-2.0E
BOT CHORD 4x2 SPF 2400F-2.0E H
WEB 4x2 HF STUD
PLATE VALUES PER ICBO RESEARCH REPORT #1607.

☒ If rigid sheathing is not directly attached to the bottom chord, adequate lateral bracing may be required (by others).
L For floor applications, 2x6 strongback bridging is recommended at 10'-0" O.C.

This design based on chord bracing applied per the following schedule:
max o.c. from to

C1

TC	FORCE	AXL	BND	CSI
1	0	.00	.16	.16
2	-3841	.14	.16	.30
3	-3841	.19	.17	.36
4	-4952	.23	.06	.29
5	-4961	.30	.17	.47
6	-4952	.23	.06	.29
7	-3841	.19	.17	.36
8	-3841	.14	.16	.30
9	0	.00	.16	.16

I

BC	FORCE	AXL	BND	CSI
1	2220	.22	.08	.30
2	4726	.47	.05	.52
3	4726	.47	.05	.52
4	4961	.49	.05	.54
5	4726	.47	.05	.52
6	2220	.22	.08	.30

WEB	FORCE	CSI	WEB	FORCE	CSI
1-11	-125	.03	6-15	-166	.04
2-11	-2414	.67	7-15	296	.13
2-12	1756	.76	7-16	-960	.27
3-12	-250	.06	8-16	-250	.06
4-12	-960	.27	9-16	1756	.76
4-14	296	.13	9-17	-2414	.67
5-14	-166	.04	10-17	-125	.03

K

Panel Lengths :
Std = 2- 5- 4
4-5 1- 2- 4 6-7 1- 2- 4

J MAX DEFLECTION (span) :
L/501 IN MEM 14-15 (LIVE)
L= -.47" D= -.17" T= -.64"

B

Joint Locations			
1	0- 0- 0	10	20- 0- 0
2	2- 8- 4	11	0- 0- 0
3	5- 2- 4	12	5- 2- 4
4	7- 8- 4	13	7- 0- 0
5	8-11- 4	14	8-11- 4
6	11- 0-12	15	11- 0-12
7	12- 3-12	16	14- 9-12
8	14- 9-12	17	20- 0- 0
9	17- 3-12		

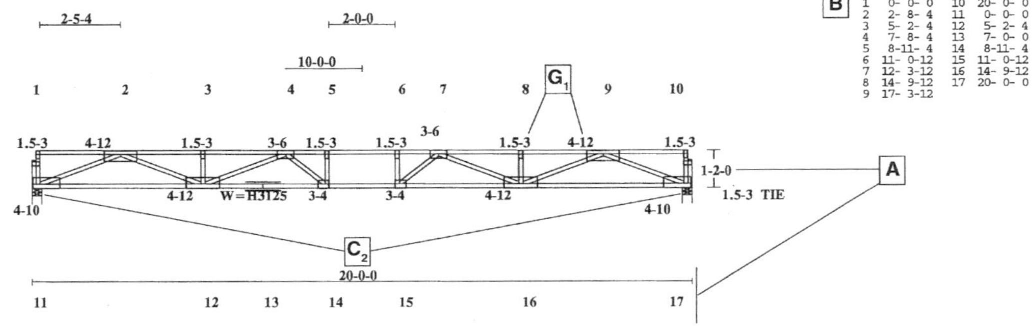

The seal on this drawing indicates acceptance of professional engineering responsibility solely for the truss component design shown. The suitability and use of this component for any particular building design is the responsibility of the building designer, per ANSI/TPI 1-1995 Section 2.

G2

All plates are 20 gauge Truswal Connectors unless preceded by "18" for 18 gauge or "H" for 16 gauge.

D

Example Created For

WTCA™

WARNING ***Read all notes on this sheet and give a copy of it to the Erecting Contractor.***
This design is for an individual building component. It has been based on specifications provided by the component manufacturer and done in accordance with the current versions of TPI and AFPA design standards. No responsibility is assumed for dimensional accuracy. Dimensions are to be verified by the component manufacturer and/or building designer prior to fabrication. The building designer shall ascertain that the loads utilized on this design meet or exceed the loading imposed by the local building code. It is assumed that the top chord is laterally braced by the roof or floor sheathing and the bottom chord is laterally braced by a rigid sheathing material directly attached, unless otherwise noted. Bracing shown is for lateral support of components members only to reduce buckling length. This component shall not be placed in any environment that will cause the moisture content of the wood exceed 19% and/or cause connector plate corrosion. Fabricate, handle, install and brace this truss in accordance with the following standards: 'TRUSCOM MANUAL', by Truswal, 'QUALITY CONTROL STANDARD FOR METAL PLATE CONNECTED WOOD TRUSSES' - (QST-88), 'HANDLING INSTALLING AND BRACING METAL PLATE CONNECTED WOOD TRUSSES' - (HIB-91) and 'HIB-91 SUMMARY SHEET' by TPI. The Truss Plate Institute (TPI) is located at 583 D'Onofrio Drive, Madison, Wisconsin 53719. The American Forest and Paper Association (AFPA) is located at 1250 Connecticut Ave, NW, Ste 200, Washington, DC 20036.

Job #: 12345		Truss: FT1
Chk:		
Dsgnr: SAK		Date: 1/14/98
TC Live	40.0 psf	DurFacs L=1.00 P=1.00
TC Dead	10.0 psf	Rep Mbr Bnd 1.15
BC Live	.0 psf	O.C.Spacing 2- 0- 0
BC Dead	5.0 psf	Design Spec UBC
TOTAL	55.0 psf	Seqn 09.15.97- 25514

E

(Courtesy: Wood Truss Council of America, Madison, Wisconsin.)

Identifying the Parts on the Truss Drawing

Key

A. Slope or depth, span and spacing [6.2.1[1]/8.2.1(a)[2]]

B. Location of all joints [6.2.2[1]/8.2.1(b)[2]]

C. Required truss bearing widths [6.2.3[1]/8.2.1(c)[2]]

D. Design loads as applicable: [6.2.4[1]/8.2.1(d)[2]]
- Top chord live load (including snow loads)
- Top chord dead load
- Bottom chord live load
- Bottom chord dead load
- Concentrated loads and their points of application
- Controlling wind and earthquake loads

E. Adjustments to lumber and metal connector plate design values for conditions of use [6.2.5[1]/8.2.1(e)[2]]

F. Each reaction force and direction [6.2.6[1]/8.2.1(f)[2]]

G. Metal connector plate type, size, thickness or gage, and the dimensioned location of each metal connector plate except where symmetrically located relative to joint interface [6.2.7[1]/8.2.1(g)[2]]

H. Lumber size, species, and grade for each member [6.2.8[1]/8.2.1(h)[2]]

I. Axial and Bending stress indices CSI (Combined Stress Index), which shows structural "efficiency" of a member, and has a maximum of 1.00

J. Calculated deflection ratio and/or maximum deflection for live and total load [6.2.10[1]]

K. Maximum axial forces in the Truss members to enable the Building Designer to design the size, connections and anchorage of the permanent continuous lateral bracing. Forces may be shown on the Truss Design Drawing or on supplemental documents [6.2.11[1]/2.2.2(e) & 5.3.3[2]]

L. Required permanent Truss member bracing location [6.2.12[1]/2.2.2(e) & 5.3.3[2]]

M. Connection requirements for:
- Truss to Truss girder
- Truss ply to ply
- Field splices

[6.2.9[1]/8.2.1(i) & 5.4.1[2] (Note: there are no examples of this on either drawing.)]

[1] Sections of WTCA 1-1995 containing more information
[2] Sections of the latest ANSI/TPI 1 containing more information

(Courtesy: Wood Truss Council of America, Madison, Wisconsin.)

Anatomy of a Tree

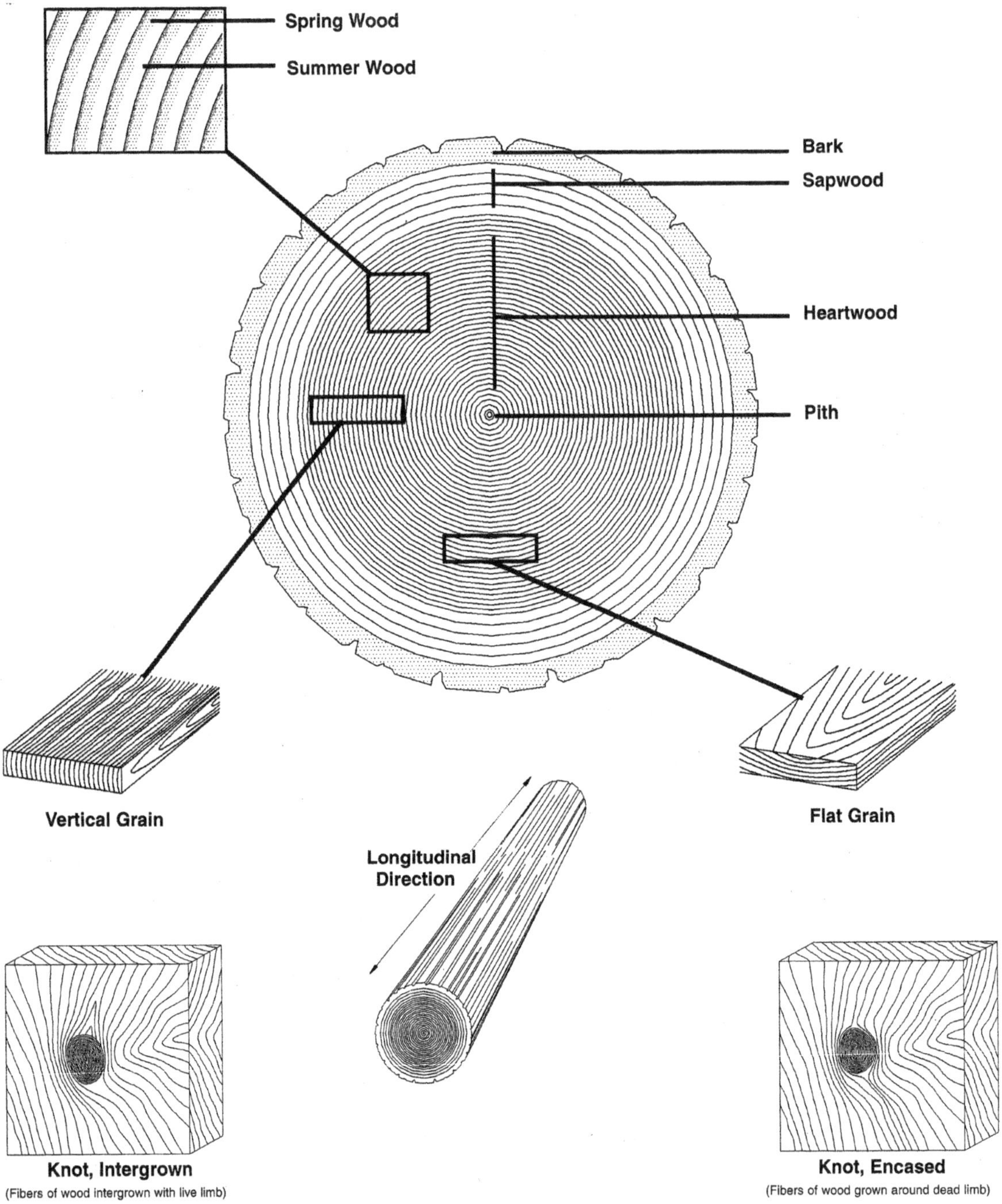

(By permission: Woodwork Institute of California, West Sacramento, California)

Types of Veneer Cuts

The manner in which veneers are cut is an important factor in producing the various visual effects obtained. Two woods of the same species, but with their veneers cut differently, will have entirely different visual character even though their color values are similar.

In plywood manufacture, the principal methods of cutting veneers are used, depending on the type of veneer required (whether for face, crossband, or core), the nature of the log, and the veneer figure desired. Primarily the veneer slicer and veneer lathe are the equipment employed. The methods are:

ROTARY

The log is mounted centrally in the lathe and turned against a razor sharp blade, like unwinding a roll of paper. Since this cut follows the log's annular growth rings, a bold variegated grain marking is produced. Rotary cut veneer is exceptionally wide.

PLAIN-SLICING (OR FLAT-SLICING)

The half log is mounted with the heart side float against the guide plate of the slicer and the slicing is done parallel to a line through the center of the log. This produces a variegated figure.

QUARTER-SLICING

The quarter log is mounted on the guide plate so that the growth rings of the log strike at approximately right angles, producing a series of stripes, straight in some woods varied in others. No limitations on the amount of medullary ray in oak.

RIFT-CUT HARDWOOD OR VERTICAL GRAIN CUT SOFTWOOD

RIFT-CUT veneer is produced in the various species of Oak. Oak has medullary ray cells which radiate from the center of the log like the curved spokes of a wheel. The rift or comb grain effect is obtained by cutting perpendicularly to these medullary rays either on the lathe or slicer. Comb grain is selected from rift. Medullary ray flake is limited.

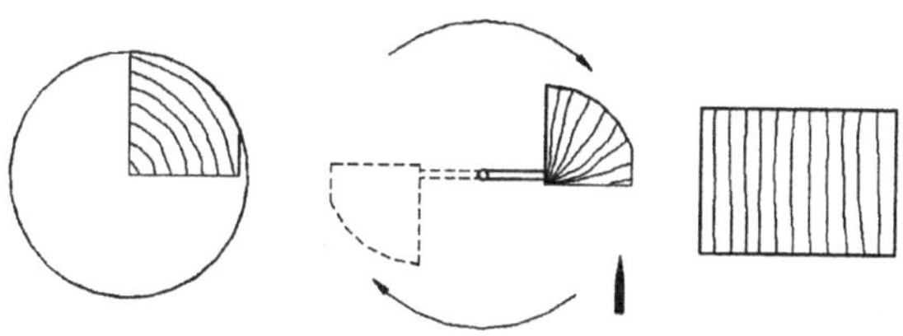

VERTICAL GRAIN cut is produced in Douglas Fir and Redwood. The vertical grain effect is produced by cutting perpendicularly to the growth rings.

(By permission: Woodwork Institute of California, West Sacramento, California)

Handing of Doors Defined and Illustrated

HANDING follows the following basic rules:

The outside of an exterior door is the key side.

The outside of an interior door is the key side or imaginary key side.

The outside of a closet door is the side opposite the closet.

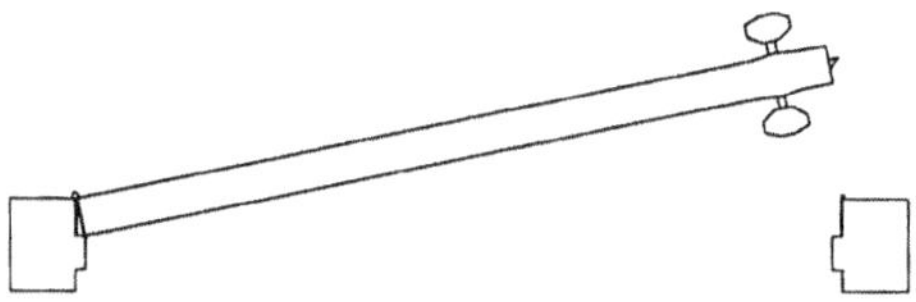

Left Hand

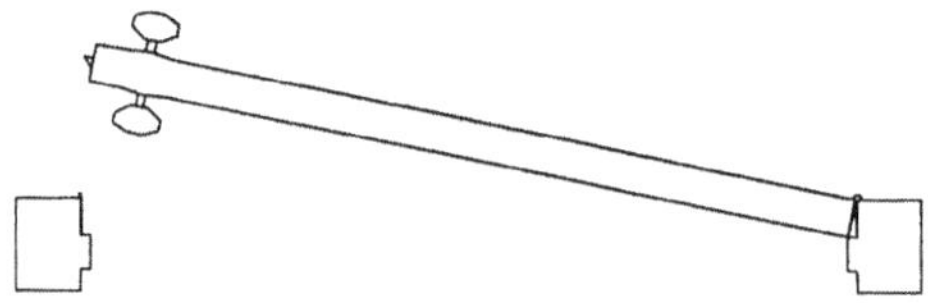

Right Hand

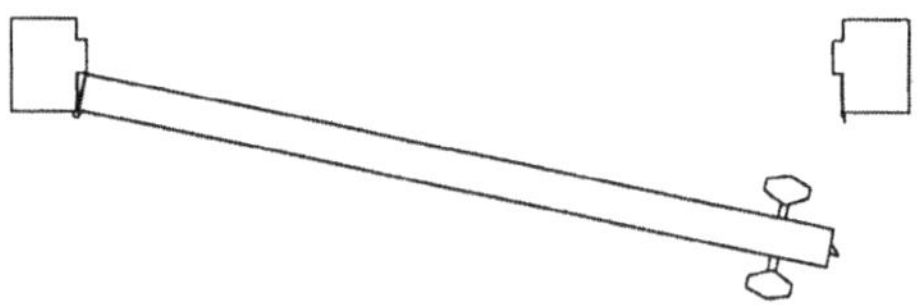

Left Hand Reversed

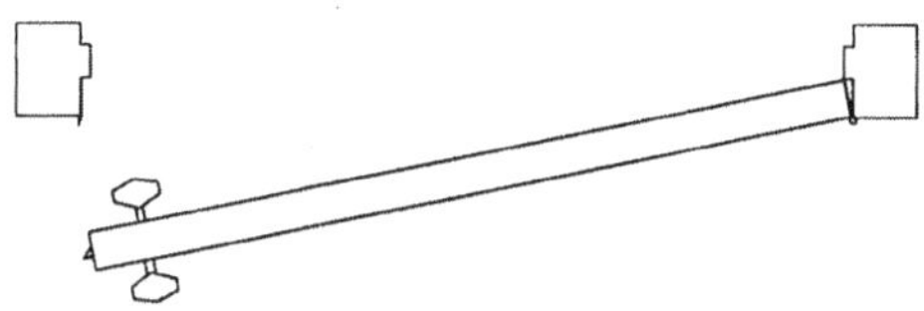

Right Hand Reversed

HAND OF DOOR IS ALWAYS DETERMINED FROM THE OUTSIDE

(By permission: Woodwork Institute of California, West Sacramento, California)

STANDARDS OF MEASUREMENT FOR LABOR PURPOSES

The tables in this guide can be used only if based on the Multiple System of measuring. In the Multiple System we base our price on the cost of painting a flat surface, such as a plain plastered wall, or a flat piece of wood or metal. All other types of surfaces are compared with painting a flat surface. Any surface that is more difficult to paint than a flat surface is multiplied by a compensating factor. For instance, a surface that takes twice as much labor is multiplied by two (2).

WINDOWS

Example of reducing a window sash to the value of a flat surface:

The multiple for this type sash is ½ or 50%. Flat surface 3x7 = 21 sq. ft. + 11 (adding multiple of 50%) = 32 sq. ft. value per side.

The multiple for this sash is qualified as follows:

Two side rails @ 7ft.	14 ft.
Two top rails @ 3ft.	6 ft.
Two middle rails @ 3ft.	6 ft.
Two vertical mullion @ 3ft.	6 ft.
Total	32 ft.
	per side

If the window sash shown were to also include jambs, frames, sill and apron, the multiple would be 3. Flat surface, including frame, would now be 3½ x 7½ = 26 sq. ft., multiplied by 3 = 72 sq. ft. per side.

The multiple for this sash and frame is qualified as follows:

Two side jambs @ 7 ft.	14 ft.
One top head @ 3 ft.	3 ft.
One sill @ 3 ft.	3 ft.
One apron @ 3 ft.	3 ft.
Two side frames @ 7 ft.	14 ft.
One top frame @ 3 ft.	3 ft.
Window sash qualified above	32 ft.
Total	72 ft.
	per side

Other windows with more lights would be higher is square foot value. The minimum value of any windows is 30 sq. ft.

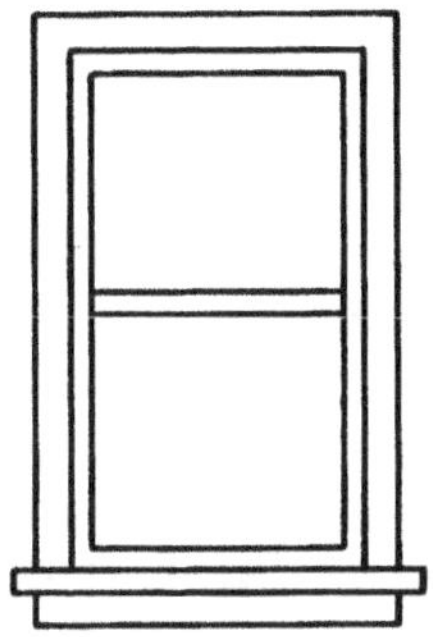

Below is a list of common surfaces requiring more labor to paint than a flat surface. The multiple or percentage generally used by estimators is as follows:

DOORS
Flush type, add 50% to area, per side.
Two-panel, add 75% to area per side.
Four-panel, double the area per side.
Six-panel, add 125% to area per side.

DOOR AND WINDOW TRIM
Linear feet plus 50% per side.
For stain and varnish finish: two member, each member is figured a foot to the foot.

CORNICE
1-story-high building, actual area x 2.
2-story-high building, actual area x 3.
1-story-high building, ornamental, actual area x 4.
2-story-high building, ornamental, actual area x 6.

SIDING, WOOD OR COMPOSITION DROP
Actual measurements (no allowance for opening of less than 100 sq. ft.) add approximately 10%.

STRUCTURAL STEEL
Actual area, add approximately 5% for rivets, bolts, plates and flanges.

OPEN TRUSSES
Length by depth by 2½.

TIE RODS
Linear feet x 2-sq. footage.

PIPES
4 in. diam.—1 ft. per foot.
4 in. to 8 in. diam.— 2 ft. per foot.
8 in. to 12 in. diam.—3 ft. per foot.
12 in. to 16 in. diam.—4 ft. per foot.
(add for pipe hangers)

CHAIN LINK FENCES
Measure on side and multiply by 3 for the two sides.

PICKET FENCES
Measure on side and multiply by 4.

GRATINGS
Measure one side, according to character, and multiply by 4 to 6.

GRILLS
(Plain) Measure one side and multiply by 2.

WAINSCOTING
If paneled, multiply actual area by 2.

BALUSTRADES
Measure one side and multiply by 4.

STAIRS
(Average) Count risers and multiply by 8 times width.

RADIATORS
For each front foot multiply the face area by 7.

CABINETS
Most 3′0″ high x 2′0″ deep base cabinets have 21 sq. ft. per 1f. to paint. Most 3′ high x 1′ deep upper cabinets have 18 sq. ft. per 1f to paint. On all other cabinets, actual sq. footage measurements should be made.

SANDING AND PUTTYING
Sanding and puttying on interior trim should be figured on high grade work at rate of 1 coat of paint. On cost of medium grade, 50% of 1 coat of paint or varnish. For industrial work, approximately 25% of the cost of 1 coat of paint or varnish.

WALL WORK
Painting Walls and Ceilings: Straight measurement over all. (No allowance for openings less than 100 sq. ft.)

OPENING ON UNIT BASIS
Most of the units on this page, such as doors, windows, cabinets, etc., can be figured on a unit basis, determining the amount of time required to paint such units on an hourly basis, or the number per eight hour basis. Some of the contractors prefer this method and find it quicker and more accurate.

Nails: Penny Designations ("d") and Lengths (U.S. and Metric)

Nail - Penny Size	Length in Inches	Length in Millimeters
2d	1	25.40
3d	1 1/4	31.75
4d	1 1/2	38.10
5d	1 3/4	44.45
6d	2	50.80
7d	2 1/4	57.15
8d	2 1/2	63.50
9d	2 3/4	69.85
10d	3	76.20
12d	3 1/4	82.55
16d	3 1/2	88.90
20d	3 3/4	95.25
30d	4 1/2	114.30
40d	5	127.00
50d	5 1/2	139.70
60d	6	152.40

Stainless Steel Sheets (Thicknesses and Weights)

Gauge	Thickness		Weight	
	Inches	Mm.	lbs/ft2	kg/m2
8	0.17188	4.3658	7.2187	44.242
10	0.14063	3.5720	5.9062	28.834
11	0.1250	3.1750	5.1500	25.6312
12	0.10938	2.7783	4.5937	22.427
14	0.07813	1.9845	3.2812	16.019
16	0.06250	1.5875	2.6250	12.815
18	0.05000	1.2700	2.1000	10.252
20	0.03750	0.9525	1.5750	7.689
22	0.03125	0.7938	1.3125	6.409
24	0.02500	0.6350	1.0500	5.126
26	0.01875	0.4763	0.7875	3.845
28	0.01563	0.3970	0.6562	3.1816
Plates				
3/16"	0.1875	4.76	7.752	37.85
1/4"	0.25	6.35	10.336	50.46
5/16"	0.3125	7.94	12.920	63.08
3/8"	0.375	9.53	15.503	75.79
1/2"	0.50	12.70	20.671	100.92
5/8"	0.625	15.88	25.839	126.15
3/4"	0.75	19.05	31.007	151.38
1"	1.00	25.4	41.342	201.83

Comparable Thicknesses and Weights of Stainless Steel, Aluminum, and Copper

STAINLESS STEEL			ALUMINUM			COPPER		
Thickness (Inch)	Gauge (U.S. Standard)	Lb/sq ft	Thickness (Inch)	Gauge (B&S)	Lb/sq ft	Thickness (Inch)	Oz sq ft	Lb/sq ft
.010	32	.420	.010	30	.141	.0108	8	.500
.0125	30	.525	.0126	28	.177	.0121	9	.563
						.0135	10	.625
.0156	28	.656	.0156		.220	.0148	11	.688
			.0179	25	.253	.0175	13	.813
.0187	26	.788						
.0219	25	.919	.020	24	.282	.021	16	1.000
.025	24	1.050	.0253	22	.352			
						.027	20	1.250
.031	22	1.313	.0313	—	.441	.032	24	1.500
.0375	20	1.575	.032	20	.451	.0337	28	1.750
			.0403	18	.563	.0431	32	2.000
			.0453	17	.100			
.050	18	2.100	.0506	16	.126			

Note that U.S. Standard Gauge (stainless sheet) is not directly comparable with the B&S Gauge (aluminum). A 20-gauge stainless averages .0375" thick; while a 20-gauge aluminum averages .032" thick; and 20-ounce copper is .027" thick. The higher strength of stainless steel permits use of thinner gauges than required for aluminum or copper, which makes stainless more competitive with aluminum on a weight-to-coverage basis and provides stainless with a substantial weight saving compared to copper. For example, 100 sq ft of .032" aluminum will weigh about 45 pounds, .021" (16-ounce) copper will weigh about 100 pounds, and .015" stainless will weigh about 66 pounds.

Wire and Sheetmetal Gauges and Weights

Name of Gage	*United States Standard Gage		The United States Steel Wire Gage	American or Brown & Sharpe Wire Gage	New Birmingham Standard Sheet & Hoop Gage	British Imperial or English Legal Standard Wire Gage	Birmingham or Stubs Iron Wire Gage	Name of Gage
Principal Use	Uncoated Steel Sheets and Light Plates		Steel Wire except Music Wire	Non-Ferrous Sheets and Wire	Iron and Steel Sheets and Hoops	Wire	Strips, Bands, Hoops and Wire	Principal Use
Gage No.	Weight Oz. per Sq. Ft.	Approx. Thickness Inches	Thickness, Inches					Gage No.
7/0's			.4900		.6666	.500		7/0's
6/0's			.4615	.5800	.625	.464		6/0's
5/0's			.4305	.5165	.5883	.432	.550	5/0's
4/0's			.3938	.4600	.5416	.400	.454	4/0's
3/0's			.3625	.3648	.500	.372	.425	3/0's
2/0's			.3310	.3249	.4452	.348	.380	2/0's
1/0			.3065	.2893	.3964	.324	.340	1/0
1			.2830	.2576	.3532	.300	.300	1
2			.2625	.2294	.3147	.276	.284	2
3	160	.2391	.2437	.2043	.2804	.252	.259	3
4	150	.2242	.2253	.1819	.250	.232	.238	4
5	140	.2092	.2070	.1620	.2225	.212	.220	5
6	130	.1943	.1920	.1443	.1981	.192	.203	6
7	120	.1793	.1770	.1285	.1764	.176	.180	7
8	110	.1644	.1620	.1144	.1570	.160	.165	8
9	100	.1495	.1483	.1019	.1398	.144	.148	9
10	90	1345	.1350	.0907	.1250	.128	.134	10
11	80	.1196	.1205	.0808	.1113	.116	.120	11
12	70	.1046	.1055	.0720	.0991	.104	.109	12
13	60	.0897	.0915	.0641	.0882	.092	.095	13
14	50	.0747	.0800	.0571	.0785	.080	.083	14
15	45	.0673	.0720	.0508	.0699	.072	.072	15
16	40	.0598	.0625	.0453	.0625	.064	.065	16
17	36	.0538	.0540	.0403	.0556	.056	.058	17
18	32	.0478	.0475	.0359	.0495	.048	.049	18
19	28	.0418	.0410	.0320	.0440	.040	.042	19
20	24	.0359	.0348	.0285	.0392	.036	.035	20
21	22	.0329	.0317	.0253	.0349	.032	.032	21
22	20	.0299	.0286	.0226	.0313	.028	.028	22
23	18	.0269	.0258	.0201	.0278	.024	.025	23
24	16	.0239	.0230	.0179	.0248	.022	.022	24
25	14	.0209	.0204	.0159	.0220	.020	.020	25
26	12	.0179	.0181	.0142	.0196	.018	.018	26
27	11	.0164	.0173	.0126	.0175	.0164	.016	27
28	10	.0149	.0162	.0113	.0156	.0148	.014	28
29	9	.0135	.0150	.0100	.0139	.0136	.013	29
30	8	.0120	.0140	.0089	.0123	.0124	.012	30
31	7	.0105	.0132	.0080	.0110	.0116	.010	31
32	6.5	.0097	.0128	.0071	.0098	.0108	.009	32
33	6	.0090	.0118	.0063	.0087	.0100	.008	33
34	5.5	.0082	.0104	.0056	.0077	.0092	.007	34
35	5	.0075	.0095	.0050	.0069	.0084	.005	35
36	4.5	.0067	.0090	.0045	.0061	.0076	.004	36
37	4.25	.0064	.0085	.0040	.0054	.0068		37
38	4	.0060	.0080	.0035	.0048	.0060		38
39			.0075	.0031	.0043	.0052		39
40			.0070		.0039	.0048		40

* U.S. Standard Gage is officially a weight gage, in oz per sq ft as tabulated. The Approx. Thickness shown is the "Manufacturers' Standard" of the American Iron and Steel Institute, based on steel as weighing 501.81 lb per cu ft (489.6 true weight plus 2.5 percent for average over-run in area and thickness).

Tap Drill Sizes for Fractional Size Threads

Approximately 65% Depth Thread / AMERICAN NATIONAL THREAD FORM

Tap Size	Threads per Inch	Hole Diameter	Drill
1/16	72	.049	3/64
1/16	64	.047	3/64
1/16	60	.046	56
5/64	72	.065	52
5/64	64	.063	1/16
5/64	60	.062	1/16
5/64	56	.061	53
3/32	60	.077	5/64
3/32	56	.076	48
3/32	50	.074	49
3/32	48	.073	49
7/64	56	.092	42
7/64	50	.090	43
7/64	48	.089	43
1/8	48	.105	36
1/8	40	.101	38
1/8	36	.098	40
1/8	32	.095	3/32
9/64	40	.116	32
9/64	36	.114	33
9/64	32	.110	35
5/32	40	.132	30
5/32	36	.129	30
5/32	32	.126	1/8
11/64	36	.145	27
11/64	32	.141	9/64
3/16	36	.161	20
3/16	32	.157	22
3/16	30	.155	23
3/16	24	.147	26
13/64	32	.173	17
13/64	30	.171	11/64
13/64	24	.163	20
7/32	32	.188	12
7/32	28	.184	13
7/32	24	.178	16
15/64	32	.204	6
15/64	28	.200	8
15/64	24	.194	10
1/4	32	.220	7/32
1/4	28	.215	3
1/4	27	.214	3
1/4	24	.209	4
1/4	20	.201	7
5/16	32	.282	9/32
5/16	27	.276	J
5/16	24	.272	I
5/16	20	.264	17/64
5/16	18	.258	F
3/8	27	.339	R
3/8	24	.334	Q
3/8	20	.326	21/64
3/8	16	.314	5/16
7/16	27	.401	Y
7/16	24	.397	X
7/16	20	.389	25/64
7/16	14	.368	U
1/2	27	.464	15/32
1/2	24	.460	29/64
1/2	20	.451	29/64
1/2	13	.425	27/64
1/2	12	.419	27/64
9/16	27	.526	17/32
9/16	18	.508	33/64
9/16	12	.481	31/64
5/8	27	.589	19/32
5/8	18	.571	37/64
5/8	12	.544	35/64
5/8	11	.536	17/32
11/16	16	.627	5/8
11/16	11	.599	19/32
3/4	27	.714	23/32
3/4	16	.689	11/16
3/4	12	.669	43/64
3/4	10	.653	21/32
13/16	12	.731	47/64
13/16	10	.715	23/32
7/8	27	.839	27/32
7/8	18	.821	53/64
7/8	14	.805	13/16
7/8	12	.794	51/64
7/8	9	.767	49/64
15/16	12	.856	55/64
15/16	9	.829	53/64
1	27	.964	31/32
1	14	.930	15/16
1	12	.919	59/64
1	8	.878	7/8
1 1/16	8	.941	15/16
1 1/8	12	1.044	1 3/64
1 1/8	7	.986	63/64
1 3/16	7	1.048	1 3/64
1 1/4	12	1.169	1 11/64
1 1/4	7	1.111	1 7/64
1 5/16	7	1.173	1 11/64
1 3/8	12	1.294	1 19/64
1 3/8	6	1.213	1 7/32
1 1/2	12	1.419	1 27/64
1 1/2	6	1.338	1 11/32
1 5/8	5 1/2	1.448	1 29/64
1 3/4	5	1.555	1 9/16
1 7/8	5	1.680	1 11/16
2	4 1/2	1.783	1 25/32
2 1/8	4 1/2	1.909	1 29/32
2 1/4	4 1/2	2.034	2 1/32
2 3/8	4	2.131	2 1/8
2 1/2	4	2.256	2 1/4
2 5/8	4	2.381	2 3/8
2 3/4	4	2.506	2 1/2
2 7/8	3 1/2	2.597	2 19/32
3	3 1/2	2.722	2 23/32
3 1/8	3 1/2	2.847	2 27/32
3 1/4	3 1/2	2.972	2 31/32
3 3/8	3 1/4	3.075	3 1/16
3 1/2	3 1/4	3.200	3 3/16
3 5/8	3 1/4	3.325	3 5/16
3 3/4	3	3.425	3 7/16
4	3	3.675	3 11/16

Common Material R-Values

R-value is a unit of measure for the rate of heat flow through a given thickness material(s) by conduction. It can include a cavity that incorporates air space reflective insulation. It is measured by the temperature difference between outside surfaces required to cause one **BTU** to flow through one square hour. A **BTU,** (British Thermal Unit), is the amount of heat required to raise temperature of one pound of water 1°F.

MATERIAL	R-value	MATERIAL	R-value	MATERIAL	R-value
1" mineral wool	3.70	3½" fiberglass	13.48	3" honeycomb	2.59
1/2" gypsum	0.45	½" mineral tile	1.19	3" isocyanurate	22.5
1/2" plywood	0.02	1" isocyanurate	7.50	3" polystyrene	12.0
1/8" floor tile	0.05	1" polystyrene	4.00	3" polyurethane	17.6
1/8" hardboard	0.09	1" wood core door	1.96	8" con. block	1.11
3/16" hardboard	0.14	6" fiberglass	19.00	insulated glass	1.65
5/8" gypsum	0.56	1" polyurethane	5.88	single glass pane	0.94

Useful Formulas

Circumference of a circle = $\pi \times$ *diameter* or 3.1416 × *diameter*
Diameter of a circle = *circumference* × 0.31831
Area of a square = *length* × *width*
Area of a rectangle = *length* × *width*
Area of a parallelogram = *base* × *perpendicular height*
Area of a triangle = $1/2$ *base* × *perpendicular weight*
Area of a circle = π *radius squared* or *diameter squared* × 0.7854
Area of an ellipse = *length* × *width* × 0.7854
Volume of a cube or rectangular prism = *length* × *width* × *height*
Volume of a triangular prism = *area of triangle* × *length*
Volume of a sphere = *diameter cubed* × 0.5236 (*diameter* × *diameter* × *diameter* × 0.5236)
Volume of a cone = $\pi \times$ *radius squared* × $1/3$ *height*
Volume of a cylinder = $\pi \times$ *radius squared* × *height*
Length of one side of a square × 1.128 = *diameter of an equal circle*
Doubling the diameter of a pipe or cylinder increases its capacity 4 times
Pressure (in lb/sq in.) *of a column of water* = *height of the column* (in feet) × 0.434
Capacity of a pipe or tank (in U.S. gallons) = *diameter squared* (in inches) × *length* (in inches) × 0.0034
1 gal water = $8\frac{1}{3}$ lb = 231 cu in.
1 cu ft water = $62\frac{1}{2}$ lb = $7\frac{1}{2}$ gal.

Decimal Equivalents of Inches in Feet and Yards

Inches	Feet	Yards
1	.0833	.0278
2	.1667	.0556
3	.2500	.0833
4	.333	.1111
5	.4166	.1389
6	.5000	.1667
7	.5833	.1944
8	.6667	.2222
9	.7500	.2500
10	.8333	.2778
11	.9166	.3056
12	1.000	.3333

Conversion of Fractions to Decimals

Fractions	Decimal	Fractions	Decimal
1/64	.015625	33/64	.515625
1/32	.03125	17/32	.53125
3/64	.046875	35/64	.546875
1/16	.0625	9/16	.5625
5/64	.078125	37/64	.578125
3/32	.09375	19/32	.59375
7/64	.109375	38/64	.609375
1/8	.125	5/8	.625
9/64	.140625	41/64	.640625
5/32	.15625	21/32	.65625
11/64	.1719	43/64	.67187
3/16	.1875	11/16	.6875
13/64	.2031	45/64	.70312
7/32	.2188	23/32	.71875
15/64	.234375	47/64	.734375
1/4	.25	3/4	.75
17/64	.265625	49/64	.765625
9/32	.28125	25/32	.78125
19/64	.296875	51/64	.796875
5/16	.3125	13/10	.8125
21/64	.328125	53/64	.828125
11/32	.34375	27/32	.84375
23/64	.359375	55/64	.859375
3/8	.375	7/8	.875
25/64	.398625	57/64	.890625
13/32	.40625	29/32	.90625
27/64	.421875	60/64	.921875
7/16	.4375	15/16	.9375
20/64	.453125	61/64	.953125
15/32	.46875	31/32	.96875
31/64	.484375	63/64	.984375
1/2	.50	1″	1.000000

Miscellaneous Tables of Weights, Measures, and Other Information

Square Measure

144 inches	1 square foot
9 square feet	1 square yard
30¼ sq. yds. 272¼ sq. ft.	1 square rod
160 square rods	1 acre
640 acres	1 square mile

Cubic Measure

1728 cubic inches	1 cubic foot
1 cubic foot	7.4805 gallons
27 cubic feet	1 cubic yard
128 cubic feet	1 cord

Dry Measure

2 pints	1 quart
8 quarts	1 peck
4 pecks	1 bushel
1 bushel	1.24 cu. feet
1 bushel	2150.42 cu inches

Liquid Measure

4 gills	1 pint
2 pints	1 quart
4 quarts	1 gallon
31½ gallons	1 barrel
2 barrels	1 hogshead

Linear Measure

12 inches	1 foot
3 feet	1 yard
16½ feet	1 rod or pole
5½ yards	1 rod or pole
40 rods or poles	1 furlong
8 furlongs	1 statute mile
320 rods	1 mile
5280 feet	1 mile
4 inches	1 hand
7.92 inches	1 link
18 inches	1 cubit
1.15156 miles	1 knot or 1 nautical mile

Weight – Avoirdupois or Commercial

437.5 grains	1 ounce
16 ounces	1 pound
112 pounds	1 hundredweight
2000 pounds	1 net ton or 1 short ton
20 hundredweight	1 gross or long ton
20 hundredweight	2240 pounds
2204.6 pounds	1 metric ton

Thermal Expansion of Various Materials

Material	Inches per inch 10^{-6} X per °F	Inches per 100′ of pipe per 100°F.	Ratio-assuming cast iron equals 1.00
Cast iron	6.2	0.745	1.00
Concrete	5.5	0.66	.89
Steel (mild)	6.5	0.780	1.05
Steel (stainless)	7.8	0.940	1.26
Copper	9.2	1.11	1.49
PVC (high impact)	55.6	6.68	8.95
ABS (type 1A)	56.2	6.75	9.05
Polyethylene (type 1)	94.5	11.4	15.30
Polyethylene (type 2)	83.3	10.0	13.40

Here is the *actual* increase in length for 50 feet of pipe and 70° temperature rise.

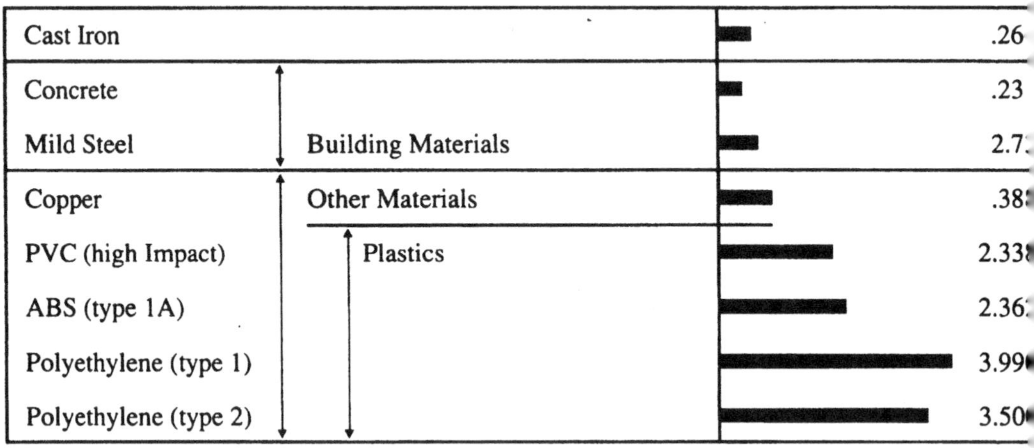

Approximate Metric Conversions

Symbol	When You Know	Multiply by	To Find	Symbol
		LENGTH		
mm	millimeters	0.04	inches	in
cm	centimeters	0.4	inches	in
m	meters	3.3	feet	ft
m	meters	1.1	yards	yd
km	kilometers	0.6	miles	mi
		AREA		
cm^2	square centimeters	0.16	square inches	in^2
m^2	square meters	1.2	square yards	yd^2
km^2	square kilometers	0.4	square miles	mi^2
ha	hectares (10,000 m^2)	2.5	acres	
		MASS (weight)		
g	grams	0.035	ounces	oz
kg	kilograms	2.2	pounds	lb
t	metric ton (1,000 kg)	1.1	short tons	
		VOLUME		
mL	milliliters	0.03	fluid ounces	fl oz
mL	milliliters	0.06	cubic inches	in^3
L	liters	2.1	pints	pt
L	liters	1.06	quarts	qt
L	liters	0.26	gallons	gal
m^3	cubic meters	35	cubic feet	ft^3
m^3	cubic meters	1.3	cubic yards	yd^3
		TEMPERATURE (exact)		
°C	degrees Celsius	multiply by 9/5, add 32	degrees Fahrenheit	°F

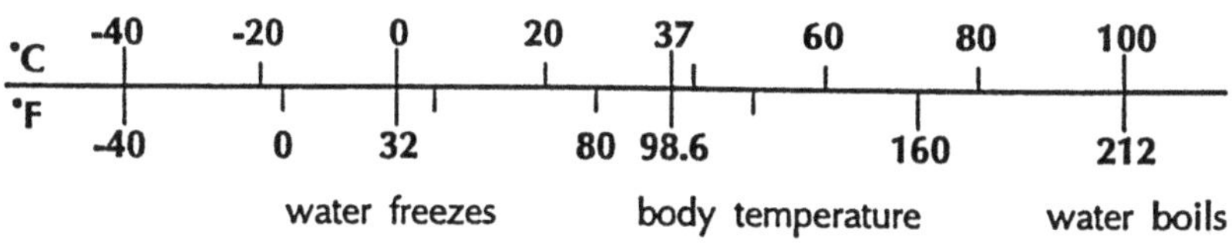

(*U.S. Department of Commerce Technology Administration, Office of Metric Programs, Washington, DC 20230*)

Metric Conversion Scales (Temperature and Measurements)

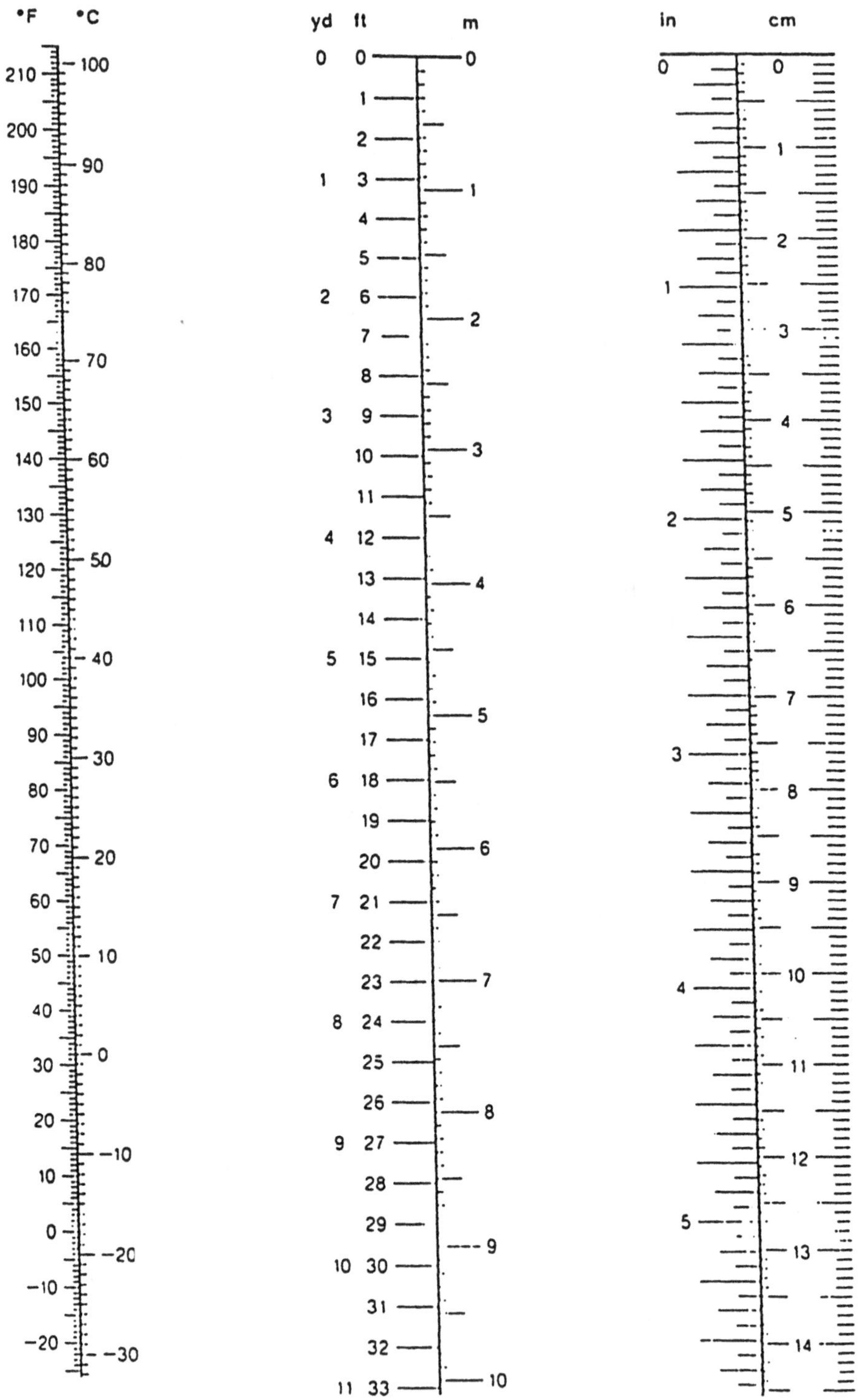

DAILY LOG

Day: ______________

January 1, 20____   **Weather: 8:00 A.M. ____ Noon: ____ 4:00 P.M. ____**

WORK FORCE

Supt.	____
Foreman	____
Carpenters	____
Laborers	____
Masons	____
Oper, Engrs.	____
Iron Wrkers	____
Electricians	____
Plumbers	____
Steam Fittrs	____
Sheet Metal	____
Glazers	____
Roofers	____
Sprinkler	____
Painters	____
Tile Setters	____
Carpet Lyrs	____
Controls	____
________	____
________	____
________	____
________	____

SUBCONTRACTORS ON PROJECT & THEIR ACTIVITY

DAILY ACTIVITY - VISITORS - INSPECTIONS

Supervisor's Signature: ______________________________

DAILY LOG

Day: ____________________

January 2, 20____   **Weather: 8:00 A.M. ___ Noon: ___ 4:00 P.M. ___**

WORK FORCE

Supt.	___
Foreman	___
Carpenters	___
Laborers	___
Masons	___
Oper, Engrs.	___
Iron Wrkers	___
Electricians	___
Plumbers	___
Steam Fittrs	___
Sheet Metal	___
Glazers	___
Roofers	___
Sprinkler	___
Painters	___
Tile Setters	___
Carpet Lyrs	___
Controls	___
________	___
________	___
________	___
________	___

SUBCONTRACTORS ON PROJECT & THEIR ACTIVITY

DAILY ACTIVITY - VISITORS - INSPECTIONS

Supervisor's Signature: ____________________________

DAILY LOG

Day: ____________________

January 3, 20____ **Weather: 8:00 A.M. ____ Noon: ____ 4:00 P.M. ____**

WORK FORCE

Supt.	____
Foreman	____
Carpenters	____
Laborers	____
Masons	____
Oper, Engrs.	____
Iron Wrkers	____
Electricians	____
Plumbers	____
Steam Fittrs	____
Sheet Metal	____
Glazers	____
Roofers	____
Sprinkler	____
Painters	____
Tile Setters	____
Carpet Lyrs	____
Controls	____
________	____
________	____
________	____
________	____

SUBCONTRACTORS ON PROJECT & THEIR ACTIVITY

DAILY ACTIVITY - VISITORS - INSPECTIONS

Supervisor's Signature: ______________________________

DAILY LOG

Day: ________________

January 4, 20___   **Weather: 8:00 A.M. ___ Noon: ___ 4:00 P.M. ___**

WORK FORCE

Supt.	___
Foreman	___
Carpenters	___
Laborers	___
Masons	___
Oper, Engrs.	___
Iron Wrkers	___
Electricians	___
Plumbers	___
Steam Fittrs	___
Sheet Metal	___
Glazers	___
Roofers	___
Sprinkler	___
Painters	___
Tile Setters	___
Carpet Lyrs	___
Controls	___
________	___
________	___
________	___
________	___

SUBCONTRACTORS ON PROJECT & THEIR ACTIVITY

DAILY ACTIVITY - VISITORS - INSPECTIONS

Supervisor's Signature: ________________________

DAILY LOG

Day: ________________

January 5, 20___ **Weather: 8:00 A.M. ___ Noon: ___ 4:00 P.M. ___**

WORK FORCE	
Supt.	___
Foreman	___
Carpenters	___
Laborers	___
Masons	___
Oper, Engrs.	___
Iron Wrkers	___
Electricians	___
Plumbers	___
Steam Fittrs	___
Sheet Metal	___
Glazers	___
Roofers	___
Sprinkler	___
Painters	___
Tile Setters	___
Carpet Lyrs	___
Controls	___
___	___
___	___
___	___
___	___

SUBCONTRACTORS ON PROJECT & THEIR ACTIVITY

DAILY ACTIVITY - VISITORS - INSPECTIONS

Supervisor's Signature: ________________

DAILY LOG

Day: ____________________

January 6, 20____ **Weather: 8:00 A.M. ___ Noon: ___ 4:00 P.M. ___**

WORK FORCE	
Supt.	____
Foreman	____
Carpenters	____
Laborers	____
Masons	____
Oper, Engrs.	____
Iron Wrkers	____
Electricians	____
Plumbers	____
Steam Fittrs	____
Sheet Metal	____
Glazers	____
Roofers	____
Sprinkler	____
Painters	____
Tile Setters	____
Carpet Lyrs	____
Controls	____
________	____
________	____
________	____
________	____

SUBCONTRACTORS ON PROJECT & THEIR ACTIVITY

DAILY ACTIVITY - VISITORS - INSPECTIONS

Supervisor's Signature: ______________________________

DAILY LOG

Day: ____________________

January 7, 20____ **Weather: 8:00 A.M. ____ Noon: ____ 4:00 P.M. ____**

WORK FORCE	
Supt.	____
Foreman	____
Carpenters	____
Laborers	____
Masons	____
Oper, Engrs.	____
Iron Wrkers	____
Electricians	____
Plumbers	____
Steam Fittrs	____
Sheet Metal	____
Glazers	____
Roofers	____
Sprinkler	____
Painters	____
Tile Setters	____
Carpet Lyrs	____
Controls	____
________	____
________	____
________	____
________	____

SUBCONTRACTORS ON PROJECT & THEIR ACTIVITY

DAILY ACTIVITY - VISITORS - INSPECTIONS

Supervisor's Signature: ____________________________

DAILY LOG

Day: ____________________

January 8, 20____ **Weather: 8:00 A.M. ___ Noon: ___ 4:00 P.M. ___**

WORK FORCE

Supt.	___
Foreman	___
Carpenters	___
Laborers	___
Masons	___
Oper, Engrs.	___
Iron Wrkers	___
Electricians	___
Plumbers	___
Steam Fittrs	___
Sheet Metal	___
Glazers	___
Roofers	___
Sprinkler	___
Painters	___
Tile Setters	___
Carpet Lyrs	___
Controls	___
______	___
______	___
______	___
______	___

SUBCONTRACTORS ON PROJECT & THEIR ACTIVITY

DAILY ACTIVITY - VISITORS - INSPECTIONS

Supervisor's Signature: ______________________________

DAILY LOG

Day: ____________________

January 9, 20____ **Weather: 8:00 A.M. ____ Noon: ____ 4:00 P.M. ____**

WORK FORCE	
Supt.	____
Foreman	____
Carpenters	____
Laborers	____
Masons	____
Oper, Engrs.	____
Iron Wrkers	____
Electricians	____
Plumbers	____
Steam Fittrs	____
Sheet Metal	____
Glazers	____
Roofers	____
Sprinkler	____
Painters	____
Tile Setters	____
Carpet Lyrs	____
Controls	____
____	____
____	____
____	____
____	____

SUBCONTRACTORS ON PROJECT & THEIR ACTIVITY

DAILY ACTIVITY - VISITORS - INSPECTIONS

Supervisor's Signature: ______________________________

DAILY LOG

Day: ____________________

January 10, 20____ **Weather: 8:00 A.M. ___ Noon: ___ 4:00 P.M. ___**

WORK FORCE	
Supt.	____
Foreman	____
Carpenters	____
Laborers	____
Masons	____
Oper, Engrs.	____
Iron Wrkers	____
Electricians	____
Plumbers	____
Steam Fittrs	____
Sheet Metal	____
Glazers	____
Roofers	____
Sprinkler	____
Painters	____
Tile Setters	____
Carpet Lyrs	____
Controls	____
________	____
________	____
________	____
________	____

SUBCONTRACTORS ON PROJECT & THEIR ACTIVITY

DAILY ACTIVITY - VISITORS - INSPECTIONS

Supervisor's Signature: ______________________________

DAILY LOG

Day: ____________________

January 11, 20____   **Weather: 8:00 A.M. ____ Noon: ____ 4:00 P.M. ____**

WORK FORCE

Supt.	____
Foreman	____
Carpenters	____
Laborers	____
Masons	____
Oper, Engrs.	____
Iron Wrkers	____
Electricians	____
Plumbers	____
Steam Fittrs	____
Sheet Metal	____
Glazers	____
Roofers	____
Sprinkler	____
Painters	____
Tile Setters	____
Carpet Lyrs	____
Controls	____
________	____
________	____
________	____
________	____

SUBCONTRACTORS ON PROJECT & THEIR ACTIVITY

DAILY ACTIVITY - VISITORS - INSPECTIONS

Supervisor's Signature: ______________________________

DAILY LOG

Day: ____________________

January 12, 20____   **Weather: 8:00 A.M. ____ Noon: ____ 4:00 P.M. ____**

WORK FORCE

Supt.	____
Foreman	____
Carpenters	____
Laborers	____
Masons	____
Oper, Engrs.	____
Iron Wrkers	____
Electricians	____
Plumbers	____
Steam Fittrs	____
Sheet Metal	____
Glazers	____
Roofers	____
Sprinkler	____
Painters	____
Tile Setters	____
Carpet Lyrs	____
Controls	____
________	____
________	____
________	____
________	____

SUBCONTRACTORS ON PROJECT & THEIR ACTIVITY

DAILY ACTIVITY - VISITORS - INSPECTIONS

Supervisor's Signature: ______________________________

DAILY LOG

Day: ____________________

January 13, 20____ **Weather: 8:00 A.M. ____ Noon: ____ 4:00 P.M. ____**

WORK FORCE

Supt.	____
Foreman	____
Carpenters	____
Laborers	____
Masons	____
Oper, Engrs.	____
Iron Wrkers	____
Electricians	____
Plumbers	____
Steam Fittrs	____
Sheet Metal	____
Glazers	____
Roofers	____
Sprinkler	____
Painters	____
Tile Setters	____
Carpet Lyrs	____
Controls	____
________	____
________	____
________	____
________	____

SUBCONTRACTORS ON PROJECT & THEIR ACTIVITY

DAILY ACTIVITY - VISITORS - INSPECTIONS

Supervisor's Signature: ______________________________

DAILY LOG

Day: ____________________

January 14, 20____   **Weather: 8:00 A.M. ___ Noon: ___ 4:00 P.M. ___**

WORK FORCE

Supt.	___
Foreman	___
Carpenters	___
Laborers	___
Masons	___
Oper, Engrs.	___
Iron Wrkers	___
Electricians	___
Plumbers	___
Steam Fittrs	___
Sheet Metal	___
Glazers	___
Roofers	___
Sprinkler	___
Painters	___
Tile Setters	___
Carpet Lyrs	___
Controls	___
________	___
________	___
________	___
________	___

SUBCONTRACTORS ON PROJECT & THEIR ACTIVITY

DAILY ACTIVITY - VISITORS - INSPECTIONS

Supervisor's Signature: ________________________________

DAILY LOG

Day: ________________

January 15, 20___ **Weather: 8:00 A.M. ___ Noon: ___ 4:00 P.M. ___**

WORK FORCE		SUBCONTRACTORS ON PROJECT & THEIR ACTIVITY
Supt.	___	
Foreman	___	
Carpenters	___	
Laborers	___	
Masons	___	
Oper, Engrs.	___	
Iron Wrkers	___	
Electricians	___	
Plumbers	___	
Steam Fittrs	___	
Sheet Metal	___	
Glazers	___	
Roofers	___	
Sprinkler	___	
Painters	___	
Tile Setters	___	
Carpet Lyrs	___	
Controls	___	
________	___	DAILY ACTIVITY - VISITORS - INSPECTIONS
________	___	
________	___	
________	___	

Supervisor's Signature: ________________________________

DAILY LOG

Day: ____________________

January 16, 20____ **Weather: 8:00 A.M. ___ Noon: ___ 4:00 P.M. ___**

WORK FORCE

Supt.	___
Foreman	___
Carpenters	___
Laborers	___
Masons	___
Oper, Engrs.	___
Iron Wrkers	___
Electricians	___
Plumbers	___
Steam Fittrs	___
Sheet Metal	___
Glazers	___
Roofers	___
Sprinkler	___
Painters	___
Tile Setters	___
Carpet Lyrs	___
Controls	___
________	___
________	___
________	___
________	___

SUBCONTRACTORS ON PROJECT & THEIR ACTIVITY

DAILY ACTIVITY - VISITORS - INSPECTIONS

Supervisor's Signature: ______________________________

DAILY LOG

Day: ________________

January 17, 20____ **Weather: 8:00 A.M. ____ Noon: ____ 4:00 P.M. ____**

WORK FORCE

Supt.	____
Foreman	____
Carpenters	____
Laborers	____
Masons	____
Oper, Engrs.	____
Iron Wrkers	____
Electricians	____
Plumbers	____
Steam Fittrs	____
Sheet Metal	____
Glazers	____
Roofers	____
Sprinkler	____
Painters	____
Tile Setters	____
Carpet Lyrs	____
Controls	____
________	____
________	____
________	____
________	____

SUBCONTRACTORS ON PROJECT & THEIR ACTIVITY

DAILY ACTIVITY - VISITORS - INSPECTIONS

Supervisor's Signature: ________________________________

DAILY LOG

Day: ____________________

January 18, 20____ **Weather: 8:00 A.M. ___ Noon: ___ 4:00 P.M. ___**

WORK FORCE

Supt.	___
Foreman	___
Carpenters	___
Laborers	___
Masons	___
Oper, Engrs.	___
Iron Wrkers	___
Electricians	___
Plumbers	___
Steam Fittrs	___
Sheet Metal	___
Glazers	___
Roofers	___
Sprinkler	___
Painters	___
Tile Setters	___
Carpet Lyrs	___
Controls	___
________	___
________	___
________	___
________	___

SUBCONTRACTORS ON PROJECT & THEIR ACTIVITY

DAILY ACTIVITY - VISITORS - INSPECTIONS

Supervisor's Signature: ____________________

DAILY LOG

Day: ________________

January 19, 20____ **Weather: 8:00 A.M. ____ Noon: ____ 4:00 P.M. ____**

WORK FORCE

Supt.	____
Foreman	____
Carpenters	____
Laborers	____
Masons	____
Oper, Engrs.	____
Iron Wrkers	____
Electricians	____
Plumbers	____
Steam Fittrs	____
Sheet Metal	____
Glazers	____
Roofers	____
Sprinkler	____
Painters	____
Tile Setters	____
Carpet Lyrs	____
Controls	____
________	____
________	____
________	____
________	____

SUBCONTRACTORS ON PROJECT & THEIR ACTIVITY

DAILY ACTIVITY - VISITORS - INSPECTIONS

Supervisor's Signature: ________________________________

DAILY LOG

Day: ____________________

January 20, 20____ **Weather: 8:00 A.M. ___ Noon: ___ 4:00 P.M. ___**

WORK FORCE

Supt.	____
Foreman	____
Carpenters	____
Laborers	____
Masons	____
Oper, Engrs.	____
Iron Wrkers	____
Electricians	____
Plumbers	____
Steam Fittrs	____
Sheet Metal	____
Glazers	____
Roofers	____
Sprinkler	____
Painters	____
Tile Setters	____
Carpet Lyrs	____
Controls	____
________	____
________	____
________	____
________	____

SUBCONTRACTORS ON PROJECT & THEIR ACTIVITY

DAILY ACTIVITY - VISITORS - INSPECTIONS

Supervisor's Signature: ________________________________

DAILY LOG

Day: ____________________

January 21, 20____ **Weather: 8:00 A.M. ___ Noon: ___ 4:00 P.M. ___**

WORK FORCE

Supt.	___
Foreman	___
Carpenters	___
Laborers	___
Masons	___
Oper, Engrs.	___
Iron Wrkers	___
Electricians	___
Plumbers	___
Steam Fittrs	___
Sheet Metal	___
Glazers	___
Roofers	___
Sprinkler	___
Painters	___
Tile Setters	___
Carpet Lyrs	___
Controls	___
______	___
______	___
______	___
______	___

SUBCONTRACTORS ON PROJECT & THEIR ACTIVITY

DAILY ACTIVITY - VISITORS - INSPECTIONS

Supervisor's Signature: ______________________________

DAILY LOG

Day: ____________________

January 22, 20____ **Weather: 8:00 A.M. ___ Noon: ___ 4:00 P.M. ___**

WORK FORCE

Supt.	____
Foreman	____
Carpenters	____
Laborers	____
Masons	____
Oper, Engrs.	____
Iron Wrkers	____
Electricians	____
Plumbers	____
Steam Fittrs	____
Sheet Metal	____
Glazers	____
Roofers	____
Sprinkler	____
Painters	____
Tile Setters	____
Carpet Lyrs	____
Controls	____
________	____
________	____
________	____
________	____

SUBCONTRACTORS ON PROJECT & THEIR ACTIVITY

DAILY ACTIVITY - VISITORS - INSPECTIONS

Supervisor's Signature: ______________________________

DAILY LOG

Day: ______________

January 23, 20___ **Weather: 8:00 A.M. ___ Noon: ___ 4:00 P.M. ___**

WORK FORCE

Supt.	___
Foreman	___
Carpenters	___
Laborers	___
Masons	___
Oper, Engrs.	___
Iron Wrkers	___
Electricians	___
Plumbers	___
Steam Fittrs	___
Sheet Metal	___
Glazers	___
Roofers	___
Sprinkler	___
Painters	___
Tile Setters	___
Carpet Lyrs	___
Controls	___
___	___
___	___
___	___
___	___

SUBCONTRACTORS ON PROJECT & THEIR ACTIVITY

DAILY ACTIVITY - VISITORS - INSPECTIONS

Supervisor's Signature: ______________________

DAILY LOG

Day: ____________________

January 24, 20____   **Weather: 8:00 A.M. ___ Noon: ___ 4:00 P.M. ___**

WORK FORCE

Supt.	___
Foreman	___
Carpenters	___
Laborers	___
Masons	___
Oper, Engrs.	___
Iron Wrkers	___
Electricians	___
Plumbers	___
Steam Fittrs	___
Sheet Metal	___
Glazers	___
Roofers	___
Sprinkler	___
Painters	___
Tile Setters	___
Carpet Lyrs	___
Controls	___
______	___
______	___
______	___
______	___

SUBCONTRACTORS ON PROJECT & THEIR ACTIVITY

DAILY ACTIVITY - VISITORS - INSPECTIONS

Supervisor's Signature: ____________________

DAILY LOG

Day: ____________________

January 25, 20____ **Weather: 8:00 A.M. ___ Noon: ___ 4:00 P.M. ___**

WORK FORCE

Supt.	___
Foreman	___
Carpenters	___
Laborers	___
Masons	___
Oper, Engrs.	___
Iron Wrkers	___
Electricians	___
Plumbers	___
Steam Fittrs	___
Sheet Metal	___
Glazers	___
Roofers	___
Sprinkler	___
Painters	___
Tile Setters	___
Carpet Lyrs	___
Controls	___
______	___
______	___
______	___
______	___

SUBCONTRACTORS ON PROJECT & THEIR ACTIVITY

DAILY ACTIVITY - VISITORS - INSPECTIONS

Supervisor's Signature: ______________________________

DAILY LOG

Day: ________________

January 26, 20___ **Weather: 8:00 A.M. ___ Noon: ___ 4:00 P.M. ___**

WORK FORCE

Supt.	___
Foreman	___
Carpenters	___
Laborers	___
Masons	___
Oper, Engrs.	___
Iron Wrkers	___
Electricians	___
Plumbers	___
Steam Fittrs	___
Sheet Metal	___
Glazers	___
Roofers	___
Sprinkler	___
Painters	___
Tile Setters	___
Carpet Lyrs	___
Controls	___
________	___
________	___
________	___
________	___

SUBCONTRACTORS ON PROJECT & THEIR ACTIVITY

DAILY ACTIVITY - VISITORS - INSPECTIONS

Supervisor's Signature: ______________________

DAILY LOG

Day: ____________________

January 27, 20____ **Weather: 8:00 A.M. ___ Noon: ___ 4:00 P.M. ___**

WORK FORCE

Supt.	___
Foreman	___
Carpenters	___
Laborers	___
Masons	___
Oper, Engrs.	___
Iron Wrkers	___
Electricians	___
Plumbers	___
Steam Fittrs	___
Sheet Metal	___
Glazers	___
Roofers	___
Sprinkler	___
Painters	___
Tile Setters	___
Carpet Lyrs	___
Controls	___
______	___
______	___
______	___
______	___

SUBCONTRACTORS ON PROJECT & THEIR ACTIVITY

DAILY ACTIVITY - VISITORS - INSPECTIONS

Supervisor's Signature: ______________________________

DAILY LOG

Day: ____________________

January 28, 20____ **Weather: 8:00 A.M. ____ Noon: ____ 4:00 P.M. ____**

WORK FORCE

Supt.	____
Foreman	____
Carpenters	____
Laborers	____
Masons	____
Oper, Engrs.	____
Iron Wrkers	____
Electricians	____
Plumbers	____
Steam Fittrs	____
Sheet Metal	____
Glazers	____
Roofers	____
Sprinkler	____
Painters	____
Tile Setters	____
Carpet Lyrs	____
Controls	____
________	____
________	____
________	____
________	____

SUBCONTRACTORS ON PROJECT & THEIR ACTIVITY

DAILY ACTIVITY - VISITORS - INSPECTIONS

Supervisor's Signature: ______________________________

DAILY LOG

Day: ____________________

January 29, 20____ **Weather: 8:00 A.M. ____ Noon: ____ 4:00 P.M. ____**

WORK FORCE	
Supt.	____
Foreman	____
Carpenters	____
Laborers	____
Masons	____
Oper, Engrs.	____
Iron Wrkers	____
Electricians	____
Plumbers	____
Steam Fittrs	____
Sheet Metal	____
Glazers	____
Roofers	____
Sprinkler	____
Painters	____
Tile Setters	____
Carpet Lyrs	____
Controls	____
________	____
________	____
________	____
________	____

SUBCONTRACTORS ON PROJECT & THEIR ACTIVITY

DAILY ACTIVITY - VISITORS - INSPECTIONS

Supervisor's Signature: ______________________________

DAILY LOG

Day: ____________________

January 30, 20____ **Weather: 8:00 A.M. ___ Noon: ___ 4:00 P.M. ___**

WORK FORCE

Supt.	___
Foreman	___
Carpenters	___
Laborers	___
Masons	___
Oper, Engrs.	___
Iron Wrkers	___
Electricians	___
Plumbers	___
Steam Fittrs	___
Sheet Metal	___
Glazers	___
Roofers	___
Sprinkler	___
Painters	___
Tile Setters	___
Carpet Lyrs	___
Controls	___
________	___
________	___
________	___
________	___

SUBCONTRACTORS ON PROJECT & THEIR ACTIVITY

DAILY ACTIVITY - VISITORS - INSPECTIONS

Supervisor's Signature: ______________________________

DAILY LOG

Day: ________________

January 31, 20____ **Weather: 8:00 A.M. ___ Noon: ___ 4:00 P.M. ___**

WORK FORCE

Supt.	___
Foreman	___
Carpenters	___
Laborers	___
Masons	___
Oper, Engrs.	___
Iron Wrkers	___
Electricians	___
Plumbers	___
Steam Fittrs	___
Sheet Metal	___
Glazers	___
Roofers	___
Sprinkler	___
Painters	___
Tile Setters	___
Carpet Lyrs	___
Controls	___
___	___
___	___
___	___
___	___

SUBCONTRACTORS ON PROJECT & THEIR ACTIVITY

DAILY ACTIVITY - VISITORS - INSPECTIONS

Supervisor's Signature: ______________________________

DAILY LOG

Day: ____________________

February 1, 20____ **Weather: 8:00 A.M. ____ Noon: ____ 4:00 P.M. ____**

WORK FORCE

Supt.	____
Foreman	____
Carpenters	____
Laborers	____
Masons	____
Oper, Engrs.	____
Iron Wrkers	____
Electricians	____
Plumbers	____
Steam Fittrs	____
Sheet Metal	____
Glazers	____
Roofers	____
Sprinkler	____
Painters	____
Tile Setters	____
Carpet Lyrs	____
Controls	____
________	____
________	____
________	____
________	____

SUBCONTRACTORS ON PROJECT & THEIR ACTIVITY

DAILY ACTIVITY - VISITORS - INSPECTIONS

Supervisor's Signature: ______________________________

DAILY LOG

Day: ____________________

February 2, 20____ **Weather: 8:00 A.M.** ____ **Noon:** ____ **4:00 P.M.** ____

WORK FORCE

Supt.	____
Foreman	____
Carpenters	____
Laborers	____
Masons	____
Oper, Engrs.	____
Iron Wrkers	____
Electricians	____
Plumbers	____
Steam Fittrs	____
Sheet Metal	____
Glazers	____
Roofers	____
Sprinkler	____
Painters	____
Tile Setters	____
Carpet Lyrs	____
Controls	____
____	____
____	____
____	____
____	____

SUBCONTRACTORS ON PROJECT & THEIR ACTIVITY

DAILY ACTIVITY - VISITORS - INSPECTIONS

Supervisor's Signature: ______________________________

DAILY LOG

Day: ____________________

February 3, 20____ **Weather: 8:00 A.M. ____ Noon: ____ 4:00 P.M. ____**

WORK FORCE

Supt.	____
Foreman	____
Carpenters	____
Laborers	____
Masons	____
Oper, Engrs.	____
Iron Wrkers	____
Electricians	____
Plumbers	____
Steam Fittrs	____
Sheet Metal	____
Glazers	____
Roofers	____
Sprinkler	____
Painters	____
Tile Setters	____
Carpet Lyrs	____
Controls	____
________	____
________	____
________	____
________	____

SUBCONTRACTORS ON PROJECT & THEIR ACTIVITY

DAILY ACTIVITY - VISITORS - INSPECTIONS

Supervisor's Signature: ______________________________

DAILY LOG

Day: ____________________

February 4, 20____ **Weather: 8:00 A.M. ____ Noon: ____ 4:00 P.M. ____**

WORK FORCE		SUBCONTRACTORS ON PROJECT & THEIR ACTIVITY
Supt.	____	
Foreman	____	
Carpenters	____	
Laborers	____	
Masons	____	
Oper, Engrs.	____	
Iron Wrkers	____	
Electricians	____	
Plumbers	____	
Steam Fittrs	____	
Sheet Metal	____	
Glazers	____	
Roofers	____	
Sprinkler	____	
Painters	____	
Tile Setters	____	
Carpet Lyrs	____	
Controls	____	
____	____	DAILY ACTIVITY - VISITORS - INSPECTIONS
____	____	
____	____	
____	____	

Supervisor's Signature: ________________________________

DAILY LOG

Day: ________________

February 5, 20___   **Weather: 8:00 A.M. ___ Noon: ___ 4:00 P.M. ___**

WORK FORCE	
Supt.	____
Foreman	____
Carpenters	____
Laborers	____
Masons	____
Oper, Engrs.	____
Iron Wrkers	____
Electricians	____
Plumbers	____
Steam Fittrs	____
Sheet Metal	____
Glazers	____
Roofers	____
Sprinkler	____
Painters	____
Tile Setters	____
Carpet Lyrs	____
Controls	____
________	____
________	____
________	____
________	____

SUBCONTRACTORS ON PROJECT & THEIR ACTIVITY

DAILY ACTIVITY - VISITORS - INSPECTIONS

Supervisor's Signature: ______________________________

DAILY LOG

Day: ________________

February 6, 20___ **Weather: 8:00 A.M. ___ Noon: ___ 4:00 P.M. ___**

WORK FORCE	
Supt.	___
Foreman	___
Carpenters	___
Laborers	___
Masons	___
Oper, Engrs.	___
Iron Wrkers	___
Electricians	___
Plumbers	___
Steam Fittrs	___
Sheet Metal	___
Glazers	___
Roofers	___
Sprinkler	___
Painters	___
Tile Setters	___
Carpet Lyrs	___
Controls	___
______	___
______	___
______	___
______	___

SUBCONTRACTORS ON PROJECT & THEIR ACTIVITY

DAILY ACTIVITY - VISITORS - INSPECTIONS

Supervisor's Signature: ______________________________

DAILY LOG

Day: ____________________

February 7, 20____ **Weather: 8:00 A.M. ___ Noon: ___ 4:00 P.M. ___**

WORK FORCE

Supt.	___
Foreman	___
Carpenters	___
Laborers	___
Masons	___
Oper, Engrs.	___
Iron Wrkers	___
Electricians	___
Plumbers	___
Steam Fittrs	___
Sheet Metal	___
Glazers	___
Roofers	___
Sprinkler	___
Painters	___
Tile Setters	___
Carpet Lyrs	___
Controls	___
______	___
______	___
______	___
______	___

SUBCONTRACTORS ON PROJECT & THEIR ACTIVITY

DAILY ACTIVITY - VISITORS - INSPECTIONS

Supervisor's Signature: ______________________________

DAILY LOG

Day: ____________________

February 8, 20____ **Weather: 8:00 A.M. ___ Noon: ___ 4:00 P.M. ___**

WORK FORCE	
Supt.	____
Foreman	____
Carpenters	____
Laborers	____
Masons	____
Oper, Engrs.	____
Iron Wrkers	____
Electricians	____
Plumbers	____
Steam Fittrs	____
Sheet Metal	____
Glazers	____
Roofers	____
Sprinkler	____
Painters	____
Tile Setters	____
Carpet Lyrs	____
Controls	____
________	____
________	____
________	____
________	____

SUBCONTRACTORS ON PROJECT & THEIR ACTIVITY

DAILY ACTIVITY - VISITORS - INSPECTIONS

Supervisor's Signature: ______________________________

DAILY LOG

Day: ____________________

February 9, 20____ **Weather: 8:00 A.M. ____ Noon: ____ 4:00 P.M. ____**

WORK FORCE

Supt.	____
Foreman	____
Carpenters	____
Laborers	____
Masons	____
Oper, Engrs.	____
Iron Wrkers	____
Electricians	____
Plumbers	____
Steam Fittrs	____
Sheet Metal	____
Glazers	____
Roofers	____
Sprinkler	____
Painters	____
Tile Setters	____
Carpet Lyrs	____
Controls	____
________	____
________	____
________	____
________	____

SUBCONTRACTORS ON PROJECT & THEIR ACTIVITY

DAILY ACTIVITY - VISITORS - INSPECTIONS

Supervisor's Signature: ____________________

DAILY LOG

Day: ____________________

February 10, 20____ **Weather: 8:00 A.M. ____ Noon: ____ 4:00 P.M. ____**

WORK FORCE		SUBCONTRACTORS ON PROJECT & THEIR ACTIVITY
Supt.	____	
Foreman	____	
Carpenters	____	
Laborers	____	
Masons	____	
Oper, Engrs.	____	
Iron Wrkers	____	
Electricians	____	
Plumbers	____	
Steam Fittrs	____	
Sheet Metal	____	
Glazers	____	
Roofers	____	
Sprinkler	____	
Painters	____	
Tile Setters	____	
Carpet Lyrs	____	
Controls	____	
____	____	DAILY ACTIVITY - VISITORS - INSPECTIONS
____	____	
____	____	
____	____	

Supervisor's Signature: ______________________________

DAILY LOG

Day: ________________

February 11, 20____ **Weather: 8:00 A.M. ___ Noon: ___ 4:00 P.M. ___**

WORK FORCE

Supt.	___
Foreman	___
Carpenters	___
Laborers	___
Masons	___
Oper, Engrs.	___
Iron Wrkers	___
Electricians	___
Plumbers	___
Steam Fittrs	___
Sheet Metal	___
Glazers	___
Roofers	___
Sprinkler	___
Painters	___
Tile Setters	___
Carpet Lyrs	___
Controls	___
___	___
___	___
___	___
___	___

SUBCONTRACTORS ON PROJECT & THEIR ACTIVITY

DAILY ACTIVITY - VISITORS - INSPECTIONS

Supervisor's Signature: ________________

DAILY LOG

Day: ________________

February 12, 20____ **Weather: 8:00 A.M. ____ Noon: ____ 4:00 P.M. ____**

WORK FORCE		**SUBCONTRACTORS ON PROJECT & THEIR ACTIVITY**
Supt.	____	
Foreman	____	
Carpenters	____	
Laborers	____	
Masons	____	
Oper, Engrs.	____	
Iron Wrkers	____	
Electricians	____	
Plumbers	____	
Steam Fittrs	____	
Sheet Metal	____	
Glazers	____	
Roofers	____	
Sprinkler	____	
Painters	____	
Tile Setters	____	
Carpet Lyrs	____	
Controls	____	
________	____	DAILY ACTIVITY - VISITORS - INSPECTIONS
________	____	
________	____	
________	____	

Supervisor's Signature: ______________________________

DAILY LOG

Day: ____________________

February 13, 20____ **Weather: 8:00 A.M. ____ Noon: ____ 4:00 P.M. ____**

WORK FORCE	
Supt.	____
Foreman	____
Carpenters	____
Laborers	____
Masons	____
Oper, Engrs.	____
Iron Wrkers	____
Electricians	____
Plumbers	____
Steam Fittrs	____
Sheet Metal	____
Glazers	____
Roofers	____
Sprinkler	____
Painters	____
Tile Setters	____
Carpet Lyrs	____
Controls	____
________	____
________	____
________	____
________	____

SUBCONTRACTORS ON PROJECT & THEIR ACTIVITY

DAILY ACTIVITY - VISITORS - INSPECTIONS

Supervisor's Signature: ______________________________

DAILY LOG

Day: ______________

February 14, 20____ **Weather: 8:00 A.M. ____ Noon: ____ 4:00 P.M. ____**

WORK FORCE	
Supt.	____
Foreman	____
Carpenters	____
Laborers	____
Masons	____
Oper, Engrs.	____
Iron Wrkers	____
Electricians	____
Plumbers	____
Steam Fittrs	____
Sheet Metal	____
Glazers	____
Roofers	____
Sprinkler	____
Painters	____
Tile Setters	____
Carpet Lyrs	____
Controls	____
____	____
____	____
____	____
____	____

SUBCONTRACTORS ON PROJECT & THEIR ACTIVITY

DAILY ACTIVITY - VISITORS - INSPECTIONS

Supervisor's Signature: ______________________________

DAILY LOG

Day: ____________________

February 15, 20____ **Weather: 8:00 A.M. ___ Noon: ___ 4:00 P.M. ___**

WORK FORCE	
Supt.	___
Foreman	___
Carpenters	___
Laborers	___
Masons	___
Oper, Engrs.	___
Iron Wrkers	___
Electricians	___
Plumbers	___
Steam Fittrs	___
Sheet Metal	___
Glazers	___
Roofers	___
Sprinkler	___
Painters	___
Tile Setters	___
Carpet Lyrs	___
Controls	___
______	___
______	___
______	___
______	___

SUBCONTRACTORS ON PROJECT & THEIR ACTIVITY

DAILY ACTIVITY - VISITORS - INSPECTIONS

Supervisor's Signature: ______________________________

DAILY LOG

Day: ____________________

February 16, 20____ **Weather: 8:00 A.M. ____ Noon: ____ 4:00 P.M. ____**

WORK FORCE	
Supt.	____
Foreman	____
Carpenters	____
Laborers	____
Masons	____
Oper, Engrs.	____
Iron Wrkers	____
Electricians	____
Plumbers	____
Steam Fittrs	____
Sheet Metal	____
Glazers	____
Roofers	____
Sprinkler	____
Painters	____
Tile Setters	____
Carpet Lyrs	____
Controls	____
________	____
________	____
________	____
________	____

SUBCONTRACTORS ON PROJECT & THEIR ACTIVITY

DAILY ACTIVITY - VISITORS - INSPECTIONS

Supervisor's Signature: ______________________________

DAILY LOG

Day: ____________________

February 17, 20____ **Weather: 8:00 A.M. ___ Noon: ___ 4:00 P.M. ___**

WORK FORCE

Supt.	___
Foreman	___
Carpenters	___
Laborers	___
Masons	___
Oper, Engrs.	___
Iron Wrkers	___
Electricians	___
Plumbers	___
Steam Fittrs	___
Sheet Metal	___
Glazers	___
Roofers	___
Sprinkler	___
Painters	___
Tile Setters	___
Carpet Lyrs	___
Controls	___
________	___
________	___
________	___
________	___

SUBCONTRACTORS ON PROJECT & THEIR ACTIVITY

DAILY ACTIVITY - VISITORS - INSPECTIONS

Supervisor's Signature: ______________________________

DAILY LOG

Day: ____________________

February 18, 20____ **Weather: 8:00 A.M. ____ Noon: ____ 4:00 P.M. ____**

WORK FORCE		SUBCONTRACTORS ON PROJECT & THEIR ACTIVITY
Supt.	____	
Foreman	____	
Carpenters	____	
Laborers	____	
Masons	____	
Oper, Engrs.	____	
Iron Wrkers	____	
Electricians	____	
Plumbers	____	
Steam Fittrs	____	
Sheet Metal	____	
Glazers	____	
Roofers	____	
Sprinkler	____	
Painters	____	
Tile Setters	____	
Carpet Lyrs	____	
Controls	____	
________	____	DAILY ACTIVITY - VISITORS - INSPECTIONS
________	____	
________	____	
________	____	

Supervisor's Signature: ______________________________

DAILY LOG

Day: ____________________

February 19, 20____ **Weather: 8:00 A.M. ____ Noon: ____ 4:00 P.M. ____**

WORK FORCE	
Supt.	____
Foreman	____
Carpenters	____
Laborers	____
Masons	____
Oper, Engrs.	____
Iron Wrkers	____
Electricians	____
Plumbers	____
Steam Fittrs	____
Sheet Metal	____
Glazers	____
Roofers	____
Sprinkler	____
Painters	____
Tile Setters	____
Carpet Lyrs	____
Controls	____
________	____
________	____
________	____
________	____

SUBCONTRACTORS ON PROJECT & THEIR ACTIVITY

DAILY ACTIVITY - VISITORS - INSPECTIONS

Supervisor's Signature: ______________________________

DAILY LOG

Day: ____________________

February 20, 20____ **Weather: 8:00 A.M. ____ Noon: ____ 4:00 P.M. ____**

WORK FORCE

Supt.	____
Foreman	____
Carpenters	____
Laborers	____
Masons	____
Oper, Engrs.	____
Iron Wrkers	____
Electricians	____
Plumbers	____
Steam Fittrs	____
Sheet Metal	____
Glazers	____
Roofers	____
Sprinkler	____
Painters	____
Tile Setters	____
Carpet Lyrs	____
Controls	____
________	____
________	____
________	____
________	____

SUBCONTRACTORS ON PROJECT & THEIR ACTIVITY

DAILY ACTIVITY - VISITORS - INSPECTIONS

Supervisor's Signature: ______________________________

DAILY LOG

Day: ____________________

February 21, 20____ **Weather: 8:00 A.M. ___ Noon: ___ 4:00 P.M. ___**

WORK FORCE

Supt.	____
Foreman	____
Carpenters	____
Laborers	____
Masons	____
Oper, Engrs.	____
Iron Wrkers	____
Electricians	____
Plumbers	____
Steam Fittrs	____
Sheet Metal	____
Glazers	____
Roofers	____
Sprinkler	____
Painters	____
Tile Setters	____
Carpet Lyrs	____
Controls	____
________	____
________	____
________	____
________	____

SUBCONTRACTORS ON PROJECT & THEIR ACTIVITY

DAILY ACTIVITY - VISITORS - INSPECTIONS

Supervisor's Signature: ______________________________

DAILY LOG

Day: ____________________

February 22, 20____ **Weather: 8:00 A.M. ___ Noon: ___ 4:00 P.M. ___**

WORK FORCE		SUBCONTRACTORS ON PROJECT & THEIR ACTIVITY
Supt.	___	
Foreman	___	
Carpenters	___	
Laborers	___	
Masons	___	
Oper, Engrs.	___	
Iron Wrkers	___	
Electricians	___	
Plumbers	___	
Steam Fittrs	___	
Sheet Metal	___	
Glazers	___	
Roofers	___	
Sprinkler	___	
Painters	___	
Tile Setters	___	
Carpet Lyrs	___	
Controls	___	
___	___	DAILY ACTIVITY - VISITORS - INSPECTIONS
___	___	
___	___	
___	___	

Supervisor's Signature: ________________________________

DAILY LOG

Day: ____________________

February 23, 20____ **Weather: 8:00 A.M. ___ Noon: ___ 4:00 P.M. ___**

WORK FORCE

Supt.	___
Foreman	___
Carpenters	___
Laborers	___
Masons	___
Oper, Engrs.	___
Iron Wrkers	___
Electricians	___
Plumbers	___
Steam Fittrs	___
Sheet Metal	___
Glazers	___
Roofers	___
Sprinkler	___
Painters	___
Tile Setters	___
Carpet Lyrs	___
Controls	___
________	___
________	___
________	___
________	___

SUBCONTRACTORS ON PROJECT & THEIR ACTIVITY

DAILY ACTIVITY - VISITORS - INSPECTIONS

Supervisor's Signature: ______________________________

DAILY LOG

Day: ____________________

February 24, 20____ **Weather: 8:00 A.M. ___ Noon: ___ 4:00 P.M. ___**

WORK FORCE

Supt.	___
Foreman	___
Carpenters	___
Laborers	___
Masons	___
Oper, Engrs.	___
Iron Wrkers	___
Electricians	___
Plumbers	___
Steam Fittrs	___
Sheet Metal	___
Glazers	___
Roofers	___
Sprinkler	___
Painters	___
Tile Setters	___
Carpet Lyrs	___
Controls	___
___	___
___	___
___	___
___	___

SUBCONTRACTORS ON PROJECT & THEIR ACTIVITY

DAILY ACTIVITY - VISITORS - INSPECTIONS

Supervisor's Signature: ______________________________

DAILY LOG

Day: ____________________

February 25, 20____ **Weather: 8:00 A.M.** ___ **Noon:** ___ **4:00 P.M.** ___

WORK FORCE

Supt.	___
Foreman	___
Carpenters	___
Laborers	___
Masons	___
Oper, Engrs.	___
Iron Wrkers	___
Electricians	___
Plumbers	___
Steam Fittrs	___
Sheet Metal	___
Glazers	___
Roofers	___
Sprinkler	___
Painters	___
Tile Setters	___
Carpet Lyrs	___
Controls	___
________	___
________	___
________	___
________	___

SUBCONTRACTORS ON PROJECT & THEIR ACTIVITY

DAILY ACTIVITY - VISITORS - INSPECTIONS

Supervisor's Signature: ______________________________

DAILY LOG

Day: ____________________

February 26, 20____ **Weather: 8:00 A.M. ____ Noon: ____ 4:00 P.M. ____**

WORK FORCE

Supt.	____
Foreman	____
Carpenters	____
Laborers	____
Masons	____
Oper, Engrs.	____
Iron Wrkers	____
Electricians	____
Plumbers	____
Steam Fittrs	____
Sheet Metal	____
Glazers	____
Roofers	____
Sprinkler	____
Painters	____
Tile Setters	____
Carpet Lyrs	____
Controls	____
________	____
________	____
________	____
________	____

SUBCONTRACTORS ON PROJECT & THEIR ACTIVITY

DAILY ACTIVITY - VISITORS - INSPECTIONS

Supervisor's Signature: ________________________________

DAILY LOG

Day: ____________________

February 27, 20____ **Weather: 8:00 A.M. ___ Noon: ___ 4:00 P.M. ___**

WORK FORCE	
Supt.	___
Foreman	___
Carpenters	___
Laborers	___
Masons	___
Oper, Engrs.	___
Iron Wrkers	___
Electricians	___
Plumbers	___
Steam Fittrs	___
Sheet Metal	___
Glazers	___
Roofers	___
Sprinkler	___
Painters	___
Tile Setters	___
Carpet Lyrs	___
Controls	___
________	___
________	___
________	___
________	___

SUBCONTRACTORS ON PROJECT & THEIR ACTIVITY

DAILY ACTIVITY - VISITORS - INSPECTIONS

Supervisor's Signature: ______________________________

DAILY LOG

Day: ________________

February 28, 20____ **Weather: 8:00 A.M. ___ Noon: ___ 4:00 P.M. ___**

WORK FORCE	
Supt.	____
Foreman	____
Carpenters	____
Laborers	____
Masons	____
Oper, Engrs.	____
Iron Wrkers	____
Electricians	____
Plumbers	____
Steam Fittrs	____
Sheet Metal	____
Glazers	____
Roofers	____
Sprinkler	____
Painters	____
Tile Setters	____
Carpet Lyrs	____
Controls	____
________	____
________	____
________	____
________	____

SUBCONTRACTORS ON PROJECT & THEIR ACTIVITY

DAILY ACTIVITY - VISITORS - INSPECTIONS

Supervisor's Signature: ________________________________

DAILY LOG

Day: ____________________

February 29, 20____　　**Weather: 8:00 A.M. ____ Noon: ____ 4:00 P.M. ____**

WORK FORCE

Supt.	____
Foreman	____
Carpenters	____
Laborers	____
Masons	____
Oper, Engrs.	____
Iron Wrkers	____
Electricians	____
Plumbers	____
Steam Fittrs	____
Sheet Metal	____
Glazers	____
Roofers	____
Sprinkler	____
Painters	____
Tile Setters	____
Carpet Lyrs	____
Controls	____
________	____
________	____
________	____
________	____

SUBCONTRACTORS ON PROJECT & THEIR ACTIVITY

DAILY ACTIVITY - VISITORS - INSPECTIONS

Supervisor's Signature: ________________________________

DAILY LOG

Day: ____________________

March 1, 20____ **Weather: 8:00 A.M. ___ Noon: ___ 4:00 P.M. ___**

WORK FORCE

Supt.	____
Foreman	____
Carpenters	____
Laborers	____
Masons	____
Oper, Engrs.	____
Iron Wrkers	____
Electricians	____
Plumbers	____
Steam Fittrs	____
Sheet Metal	____
Glazers	____
Roofers	____
Sprinkler	____
Painters	____
Tile Setters	____
Carpet Lyrs	____
Controls	____
________	____
________	____
________	____
________	____

SUBCONTRACTORS ON PROJECT & THEIR ACTIVITY

DAILY ACTIVITY - VISITORS - INSPECTIONS

Supervisor's Signature: ______________________________

DAILY LOG

Day: ____________________

March 2, 20____ **Weather: 8:00 A.M. ___ Noon: ___ 4:00 P.M. ___**

WORK FORCE

Supt.	____
Foreman	____
Carpenters	____
Laborers	____
Masons	____
Oper, Engrs.	____
Iron Wrkers	____
Electricians	____
Plumbers	____
Steam Fittrs	____
Sheet Metal	____
Glazers	____
Roofers	____
Sprinkler	____
Painters	____
Tile Setters	____
Carpet Lyrs	____
Controls	____
________	____
________	____
________	____
________	____

SUBCONTRACTORS ON PROJECT & THEIR ACTIVITY

DAILY ACTIVITY - VISITORS - INSPECTIONS

Supervisor's Signature: ______________________________

DAILY LOG

Day: ________________

March 3, 20___   **Weather: 8:00 A.M. ___ Noon: ___ 4:00 P.M. ___**

WORK FORCE	
Supt.	___
Foreman	___
Carpenters	___
Laborers	___
Masons	___
Oper, Engrs.	___
Iron Wrkers	___
Electricians	___
Plumbers	___
Steam Fittrs	___
Sheet Metal	___
Glazers	___
Roofers	___
Sprinkler	___
Painters	___
Tile Setters	___
Carpet Lyrs	___
Controls	___
___	___
___	___
___	___
___	___

SUBCONTRACTORS ON PROJECT & THEIR ACTIVITY

DAILY ACTIVITY - VISITORS - INSPECTIONS

Supervisor's Signature: ________________________________

DAILY LOG

Day: ____________________

March 4, 20____ **Weather: 8:00 A.M. ____ Noon: ____ 4:00 P.M. ____**

WORK FORCE

Supt.	____
Foreman	____
Carpenters	____
Laborers	____
Masons	____
Oper, Engrs.	____
Iron Wrkers	____
Electricians	____
Plumbers	____
Steam Fittrs	____
Sheet Metal	____
Glazers	____
Roofers	____
Sprinkler	____
Painters	____
Tile Setters	____
Carpet Lyrs	____
Controls	____
________	____
________	____
________	____
________	____

SUBCONTRACTORS ON PROJECT & THEIR ACTIVITY

DAILY ACTIVITY - VISITORS - INSPECTIONS

Supervisor's Signature: ____________________

DAILY LOG

Day: ____________________

March 5, 20____ **Weather: 8:00 A.M. ____ Noon: ____ 4:00 P.M. ____**

WORK FORCE

Supt.	____
Foreman	____
Carpenters	____
Laborers	____
Masons	____
Oper, Engrs.	____
Iron Wrkers	____
Electricians	____
Plumbers	____
Steam Fittrs	____
Sheet Metal	____
Glazers	____
Roofers	____
Sprinkler	____
Painters	____
Tile Setters	____
Carpet Lyrs	____
Controls	____
________	____
________	____
________	____
________	____

SUBCONTRACTORS ON PROJECT & THEIR ACTIVITY

DAILY ACTIVITY - VISITORS - INSPECTIONS

Supervisor's Signature: ________________________________

DAILY LOG

Day: ____________________

March 6, 20____ **Weather: 8:00 A.M. ____ Noon: ____ 4:00 P.M. ____**

WORK FORCE

Supt.	____
Foreman	____
Carpenters	____
Laborers	____
Masons	____
Oper, Engrs.	____
Iron Wrkers	____
Electricians	____
Plumbers	____
Steam Fittrs	____
Sheet Metal	____
Glazers	____
Roofers	____
Sprinkler	____
Painters	____
Tile Setters	____
Carpet Lyrs	____
Controls	____
________	____
________	____
________	____
________	____

SUBCONTRACTORS ON PROJECT & THEIR ACTIVITY

DAILY ACTIVITY - VISITORS - INSPECTIONS

Supervisor's Signature: ______________________________

DAILY LOG

Day: ____________________

March 7, 20____

Weather: 8:00 A.M. ____ Noon: ____ 4:00 P.M. ____

WORK FORCE	
Supt.	____
Foreman	____
Carpenters	____
Laborers	____
Masons	____
Oper, Engrs.	____
Iron Wrkers	____
Electricians	____
Plumbers	____
Steam Fittrs	____
Sheet Metal	____
Glazers	____
Roofers	____
Sprinkler	____
Painters	____
Tile Setters	____
Carpet Lyrs	____
Controls	____
________	____
________	____
________	____
________	____

SUBCONTRACTORS ON PROJECT & THEIR ACTIVITY

DAILY ACTIVITY - VISITORS - INSPECTIONS

Supervisor's Signature: ________________________________

DAILY LOG

Day: ____________________

March 8, 20____

Weather: 8:00 A.M. ____ Noon: ____ 4:00 P.M. ____

WORK FORCE

Supt.	____
Foreman	____
Carpenters	____
Laborers	____
Masons	____
Oper, Engrs.	____
Iron Wrkers	____
Electricians	____
Plumbers	____
Steam Fittrs	____
Sheet Metal	____
Glazers	____
Roofers	____
Sprinkler	____
Painters	____
Tile Setters	____
Carpet Lyrs	____
Controls	____
________	____
________	____
________	____
________	____

SUBCONTRACTORS ON PROJECT & THEIR ACTIVITY

DAILY ACTIVITY - VISITORS - INSPECTIONS

Supervisor's Signature: ______________________________

DAILY LOG

Day: ________________

March 9, 20____ **Weather: 8:00 A.M. ____ Noon: ____ 4:00 P.M. ____**

WORK FORCE

Supt.	____
Foreman	____
Carpenters	____
Laborers	____
Masons	____
Oper, Engrs.	____
Iron Wrkers	____
Electricians	____
Plumbers	____
Steam Fittrs	____
Sheet Metal	____
Glazers	____
Roofers	____
Sprinkler	____
Painters	____
Tile Setters	____
Carpet Lyrs	____
Controls	____
____	____
____	____
____	____
____	____

SUBCONTRACTORS ON PROJECT & THEIR ACTIVITY

DAILY ACTIVITY - VISITORS - INSPECTIONS

Supervisor's Signature: ______________________________

DAILY LOG

Day: ____________________

March 10, 20___   **Weather: 8:00 A.M. ___ Noon: ___ 4:00 P.M. ___**

WORK FORCE

Supt.	___
Foreman	___
Carpenters	___
Laborers	___
Masons	___
Oper, Engrs.	___
Iron Wrkers	___
Electricians	___
Plumbers	___
Steam Fittrs	___
Sheet Metal	___
Glazers	___
Roofers	___
Sprinkler	___
Painters	___
Tile Setters	___
Carpet Lyrs	___
Controls	___
___	___
___	___
___	___
___	___

SUBCONTRACTORS ON PROJECT & THEIR ACTIVITY

DAILY ACTIVITY - VISITORS - INSPECTIONS

Supervisor's Signature: ____________________

DAILY LOG

Day: ____________________

March 11, 20____ **Weather: 8:00 A.M. ____ Noon: ____ 4:00 P.M. ____**

WORK FORCE

Supt.	____
Foreman	____
Carpenters	____
Laborers	____
Masons	____
Oper, Engrs.	____
Iron Wrkers	____
Electricians	____
Plumbers	____
Steam Fittrs	____
Sheet Metal	____
Glazers	____
Roofers	____
Sprinkler	____
Painters	____
Tile Setters	____
Carpet Lyrs	____
Controls	____
________	____
________	____
________	____
________	____

SUBCONTRACTORS ON PROJECT & THEIR ACTIVITY

DAILY ACTIVITY - VISITORS - INSPECTIONS

Supervisor's Signature: ______________________________

DAILY LOG

Day: ________________

March 12, 20___ **Weather: 8:00 A.M. ___ Noon: ___ 4:00 P.M. ___**

WORK FORCE

Supt.	___
Foreman	___
Carpenters	___
Laborers	___
Masons	___
Oper, Engrs.	___
Iron Wrkers	___
Electricians	___
Plumbers	___
Steam Fittrs	___
Sheet Metal	___
Glazers	___
Roofers	___
Sprinkler	___
Painters	___
Tile Setters	___
Carpet Lyrs	___
Controls	___
______	___
______	___
______	___
______	___

SUBCONTRACTORS ON PROJECT & THEIR ACTIVITY

DAILY ACTIVITY - VISITORS - INSPECTIONS

Supervisor's Signature: ______________________

DAILY LOG

Day: ____________________

March 13, 20____ **Weather: 8:00 A.M. ____ Noon: ____ 4:00 P.M. ____**

WORK FORCE	
Supt.	____
Foreman	____
Carpenters	____
Laborers	____
Masons	____
Oper, Engrs.	____
Iron Wrkers	____
Electricians	____
Plumbers	____
Steam Fittrs	____
Sheet Metal	____
Glazers	____
Roofers	____
Sprinkler	____
Painters	____
Tile Setters	____
Carpet Lyrs	____
Controls	____
________	____
________	____
________	____
________	____

SUBCONTRACTORS ON PROJECT & THEIR ACTIVITY

DAILY ACTIVITY - VISITORS - INSPECTIONS

Supervisor's Signature: ______________________________

DAILY LOG

Day: ____________________

March 14, 20____ **Weather: 8:00 A.M. ___ Noon: ___ 4:00 P.M. ___**

WORK FORCE	
Supt.	___
Foreman	___
Carpenters	___
Laborers	___
Masons	___
Oper, Engrs.	___
Iron Wrkers	___
Electricians	___
Plumbers	___
Steam Fittrs	___
Sheet Metal	___
Glazers	___
Roofers	___
Sprinkler	___
Painters	___
Tile Setters	___
Carpet Lyrs	___
Controls	___
________	___
________	___
________	___
________	___

SUBCONTRACTORS ON PROJECT & THEIR ACTIVITY

DAILY ACTIVITY - VISITORS - INSPECTIONS

Supervisor's Signature: ______________________________

DAILY LOG

Day: ____________________

March 15, 20____ **Weather: 8:00 A.M. ____ Noon: ____ 4:00 P.M. ____**

WORK FORCE

Supt.	____
Foreman	____
Carpenters	____
Laborers	____
Masons	____
Oper, Engrs.	____
Iron Wrkers	____
Electricians	____
Plumbers	____
Steam Fittrs	____
Sheet Metal	____
Glazers	____
Roofers	____
Sprinkler	____
Painters	____
Tile Setters	____
Carpet Lyrs	____
Controls	____
________	____
________	____
________	____
________	____

SUBCONTRACTORS ON PROJECT & THEIR ACTIVITY

DAILY ACTIVITY - VISITORS - INSPECTIONS

Supervisor's Signature: ________________________________

DAILY LOG

Day: ______________

March 16, 20___ **Weather: 8:00 A.M.** ___ **Noon:** ___ **4:00 P.M.** ___

WORK FORCE

Supt.	___
Foreman	___
Carpenters	___
Laborers	___
Masons	___
Oper, Engrs.	___
Iron Wrkers	___
Electricians	___
Plumbers	___
Steam Fittrs	___
Sheet Metal	___
Glazers	___
Roofers	___
Sprinkler	___
Painters	___
Tile Setters	___
Carpet Lyrs	___
Controls	___
___	___
___	___
___	___
___	___

SUBCONTRACTORS ON PROJECT & THEIR ACTIVITY

DAILY ACTIVITY - VISITORS - INSPECTIONS

Supervisor's Signature: ______________

DAILY LOG

Day: ____________________

March 17, 20____ **Weather: 8:00 A.M. ____ Noon: ____ 4:00 P.M. ____**

WORK FORCE

Supt.	____
Foreman	____
Carpenters	____
Laborers	____
Masons	____
Oper, Engrs.	____
Iron Wrkers	____
Electricians	____
Plumbers	____
Steam Fittrs	____
Sheet Metal	____
Glazers	____
Roofers	____
Sprinkler	____
Painters	____
Tile Setters	____
Carpet Lyrs	____
Controls	____
________	____
________	____
________	____
________	____

SUBCONTRACTORS ON PROJECT & THEIR ACTIVITY

DAILY ACTIVITY - VISITORS - INSPECTIONS

Supervisor's Signature: ________________________________

DAILY LOG

Day: ____________________

March 18, 20____ **Weather: 8:00 A.M. ___ Noon: ___ 4:00 P.M. ___**

WORK FORCE

Supt.	___
Foreman	___
Carpenters	___
Laborers	___
Masons	___
Oper, Engrs.	___
Iron Wrkers	___
Electricians	___
Plumbers	___
Steam Fittrs	___
Sheet Metal	___
Glazers	___
Roofers	___
Sprinkler	___
Painters	___
Tile Setters	___
Carpet Lyrs	___
Controls	___
________	___
________	___
________	___
________	___

SUBCONTRACTORS ON PROJECT & THEIR ACTIVITY

DAILY ACTIVITY - VISITORS - INSPECTIONS

Supervisor's Signature: ______________________________

DAILY LOG

Day: ____________________

March 19, 20____ **Weather: 8:00 A.M. ____ Noon: ____ 4:00 P.M. ____**

WORK FORCE

Supt.	____
Foreman	____
Carpenters	____
Laborers	____
Masons	____
Oper, Engrs.	____
Iron Wrkers	____
Electricians	____
Plumbers	____
Steam Fittrs	____
Sheet Metal	____
Glazers	____
Roofers	____
Sprinkler	____
Painters	____
Tile Setters	____
Carpet Lyrs	____
Controls	____
________	____
________	____
________	____
________	____

SUBCONTRACTORS ON PROJECT & THEIR ACTIVITY

DAILY ACTIVITY - VISITORS - INSPECTIONS

Supervisor's Signature: ________________________________

DAILY LOG

Day: ____________________

March 20, 20____ **Weather: 8:00 A.M. ___ Noon: ___ 4:00 P.M. ___**

WORK FORCE	
Supt.	____
Foreman	____
Carpenters	____
Laborers	____
Masons	____
Oper, Engrs.	____
Iron Wrkers	____
Electricians	____
Plumbers	____
Steam Fittrs	____
Sheet Metal	____
Glazers	____
Roofers	____
Sprinkler	____
Painters	____
Tile Setters	____
Carpet Lyrs	____
Controls	____
________	____
________	____
________	____
________	____

SUBCONTRACTORS ON PROJECT & THEIR ACTIVITY

DAILY ACTIVITY - VISITORS - INSPECTIONS

Supervisor's Signature: ______________________________

DAILY LOG

Day: ____________________

March 21, 20____ **Weather: 8:00 A.M. ____ Noon: ____ 4:00 P.M. ____**

WORK FORCE

Supt.	____
Foreman	____
Carpenters	____
Laborers	____
Masons	____
Oper, Engrs.	____
Iron Wrkers	____
Electricians	____
Plumbers	____
Steam Fittrs	____
Sheet Metal	____
Glazers	____
Roofers	____
Sprinkler	____
Painters	____
Tile Setters	____
Carpet Lyrs	____
Controls	____
________	____
________	____
________	____
________	____

SUBCONTRACTORS ON PROJECT & THEIR ACTIVITY

____

DAILY ACTIVITY - VISITORS - INSPECTIONS

____

Supervisor's Signature: ____

DAILY LOG

Day: ____________________

March 22, 20____ **Weather: 8:00 A.M. ___ Noon: ___ 4:00 P.M. ___**

WORK FORCE

Supt.	___
Foreman	___
Carpenters	___
Laborers	___
Masons	___
Oper, Engrs.	___
Iron Wrkers	___
Electricians	___
Plumbers	___
Steam Fittrs	___
Sheet Metal	___
Glazers	___
Roofers	___
Sprinkler	___
Painters	___
Tile Setters	___
Carpet Lyrs	___
Controls	___
________	___
________	___
________	___
________	___

SUBCONTRACTORS ON PROJECT & THEIR ACTIVITY

DAILY ACTIVITY - VISITORS - INSPECTIONS

Supervisor's Signature: ________________________________

DAILY LOG

Day: ____________________

March 23, 20____ **Weather: 8:00 A.M. ____ Noon: ____ 4:00 P.M. ____**

WORK FORCE

Supt.	____
Foreman	____
Carpenters	____
Laborers	____
Masons	____
Oper, Engrs.	____
Iron Wrkers	____
Electricians	____
Plumbers	____
Steam Fittrs	____
Sheet Metal	____
Glazers	____
Roofers	____
Sprinkler	____
Painters	____
Tile Setters	____
Carpet Lyrs	____
Controls	____
________	____
________	____
________	____
________	____

SUBCONTRACTORS ON PROJECT & THEIR ACTIVITY

DAILY ACTIVITY - VISITORS - INSPECTIONS

Supervisor's Signature: ____

DAILY LOG

Day: ________________

March 24, 20___ **Weather: 8:00 A.M. ___ Noon: ___ 4:00 P.M. ___**

WORK FORCE	
Supt.	___
Foreman	___
Carpenters	___
Laborers	___
Masons	___
Oper, Engrs.	___
Iron Wrkers	___
Electricians	___
Plumbers	___
Steam Fittrs	___
Sheet Metal	___
Glazers	___
Roofers	___
Sprinkler	___
Painters	___
Tile Setters	___
Carpet Lyrs	___
Controls	___
______	___
______	___
______	___
______	___

SUBCONTRACTORS ON PROJECT & THEIR ACTIVITY

DAILY ACTIVITY - VISITORS - INSPECTIONS

Supervisor's Signature: ______________________

DAILY LOG

Day: ________________

March 25, 20____ **Weather: 8:00 A.M. ___ Noon: ___ 4:00 P.M. ___**

WORK FORCE

Supt.	____
Foreman	____
Carpenters	____
Laborers	____
Masons	____
Oper, Engrs.	____
Iron Wrkers	____
Electricians	____
Plumbers	____
Steam Fittrs	____
Sheet Metal	____
Glazers	____
Roofers	____
Sprinkler	____
Painters	____
Tile Setters	____
Carpet Lyrs	____
Controls	____
____	____
____	____
____	____
____	____

SUBCONTRACTORS ON PROJECT & THEIR ACTIVITY

DAILY ACTIVITY - VISITORS - INSPECTIONS

Supervisor's Signature: ____________________

DAILY LOG

Day: ____________________

March 26, 20____ **Weather: 8:00 A.M. ___ Noon: ___ 4:00 P.M. ___**

WORK FORCE		**SUBCONTRACTORS ON PROJECT & THEIR ACTIVITY**
Supt.	____	
Foreman	____	
Carpenters	____	
Laborers	____	
Masons	____	
Oper, Engrs.	____	
Iron Wrkers	____	
Electricians	____	
Plumbers	____	
Steam Fittrs	____	
Sheet Metal	____	
Glazers	____	
Roofers	____	
Sprinkler	____	
Painters	____	
Tile Setters	____	
Carpet Lyrs	____	
Controls	____	
________	____	DAILY ACTIVITY - VISITORS - INSPECTIONS
________	____	
________	____	
________	____	

Supervisor's Signature: ______________________________

DAILY LOG

Day: ____________________

March 27, 20____ **Weather: 8:00 A.M. ___ Noon: ___ 4:00 P.M. ___**

WORK FORCE

Supt.	___
Foreman	___
Carpenters	___
Laborers	___
Masons	___
Oper, Engrs.	___
Iron Wrkers	___
Electricians	___
Plumbers	___
Steam Fittrs	___
Sheet Metal	___
Glazers	___
Roofers	___
Sprinkler	___
Painters	___
Tile Setters	___
Carpet Lyrs	___
Controls	___
______	___
______	___
______	___
______	___

SUBCONTRACTORS ON PROJECT & THEIR ACTIVITY

DAILY ACTIVITY - VISITORS - INSPECTIONS

Supervisor's Signature: ______________________________

DAILY LOG

Day: ____________________

March 28, 20____ **Weather: 8:00 A.M. ___ Noon: ___ 4:00 P.M. ___**

WORK FORCE		SUBCONTRACTORS ON PROJECT & THEIR ACTIVITY
Supt.	____	
Foreman	____	
Carpenters	____	
Laborers	____	
Masons	____	
Oper, Engrs.	____	
Iron Wrkers	____	
Electricians	____	
Plumbers	____	
Steam Fittrs	____	
Sheet Metal	____	
Glazers	____	
Roofers	____	
Sprinkler	____	
Painters	____	
Tile Setters	____	
Carpet Lyrs	____	
Controls	____	
________	____	DAILY ACTIVITY - VISITORS - INSPECTIONS
________	____	
________	____	
________	____	

Supervisor's Signature: ________________________________

DAILY LOG

Day: ________________

March 29, 20___ **Weather: 8:00 A.M. ___ Noon: ___ 4:00 P.M. ___**

WORK FORCE		SUBCONTRACTORS ON PROJECT & THEIR ACTIVITY
Supt.	___	
Foreman	___	
Carpenters	___	
Laborers	___	
Masons	___	
Oper, Engrs.	___	
Iron Wrkers	___	
Electricians	___	
Plumbers	___	
Steam Fittrs	___	
Sheet Metal	___	
Glazers	___	
Roofers	___	
Sprinkler	___	
Painters	___	
Tile Setters	___	
Carpet Lyrs	___	
Controls	___	
___	___	DAILY ACTIVITY - VISITORS - INSPECTIONS
___	___	
___	___	
___	___	

Supervisor's Signature: ______________________________

DAILY LOG

Day: ____________________

March 30, 20____ **Weather: 8:00 A.M. ____ Noon: ____ 4:00 P.M. ____**

WORK FORCE

Supt.	____
Foreman	____
Carpenters	____
Laborers	____
Masons	____
Oper, Engrs.	____
Iron Wrkers	____
Electricians	____
Plumbers	____
Steam Fittrs	____
Sheet Metal	____
Glazers	____
Roofers	____
Sprinkler	____
Painters	____
Tile Setters	____
Carpet Lyrs	____
Controls	____
________	____
________	____
________	____
________	____

SUBCONTRACTORS ON PROJECT & THEIR ACTIVITY

DAILY ACTIVITY - VISITORS - INSPECTIONS

Supervisor's Signature: ______________________________

DAILY LOG

Day: ________________

March 31, 20___

Weather: 8:00 A.M. ___ Noon: ___ 4:00 P.M. ___

WORK FORCE

Supt.	___
Foreman	___
Carpenters	___
Laborers	___
Masons	___
Oper, Engrs.	___
Iron Wrkers	___
Electricians	___
Plumbers	___
Steam Fittrs	___
Sheet Metal	___
Glazers	___
Roofers	___
Sprinkler	___
Painters	___
Tile Setters	___
Carpet Lyrs	___
Controls	___
___	___
___	___
___	___
___	___

SUBCONTRACTORS ON PROJECT & THEIR ACTIVITY

DAILY ACTIVITY - VISITORS - INSPECTIONS

Supervisor's Signature: ____________________

DAILY LOG

Day: ________________

April 1, 20____ **Weather: 8:00 A.M. ____ Noon: ____ 4:00 P.M. ____**

WORK FORCE	
Supt.	____
Foreman	____
Carpenters	____
Laborers	____
Masons	____
Oper, Engrs.	____
Iron Wrkers	____
Electricians	____
Plumbers	____
Steam Fittrs	____
Sheet Metal	____
Glazers	____
Roofers	____
Sprinkler	____
Painters	____
Tile Setters	____
Carpet Lyrs	____
Controls	____
________	____
________	____
________	____
________	____

SUBCONTRACTORS ON PROJECT & THEIR ACTIVITY

DAILY ACTIVITY - VISITORS - INSPECTIONS

Supervisor's Signature: ________________________________

DAILY LOG

Day: ________________

April 2, 20____ **Weather: 8:00 A.M. ____ Noon: ____ 4:00 P.M. ____**

WORK FORCE

Supt.	____
Foreman	____
Carpenters	____
Laborers	____
Masons	____
Oper, Engrs.	____
Iron Wrkers	____
Electricians	____
Plumbers	____
Steam Fittrs	____
Sheet Metal	____
Glazers	____
Roofers	____
Sprinkler	____
Painters	____
Tile Setters	____
Carpet Lyrs	____
Controls	____
________	____
________	____
________	____
________	____

SUBCONTRACTORS ON PROJECT & THEIR ACTIVITY

DAILY ACTIVITY - VISITORS - INSPECTIONS

Supervisor's Signature: ________________________________

DAILY LOG

Day: ____________________

April 3, 20____

Weather: 8:00 A.M. ___ Noon: ___ 4:00 P.M. ___

WORK FORCE	
Supt.	___
Foreman	___
Carpenters	___
Laborers	___
Masons	___
Oper, Engrs.	___
Iron Wrkers	___
Electricians	___
Plumbers	___
Steam Fittrs	___
Sheet Metal	___
Glazers	___
Roofers	___
Sprinkler	___
Painters	___
Tile Setters	___
Carpet Lyrs	___
Controls	___
______	___
______	___
______	___
______	___

SUBCONTRACTORS ON PROJECT & THEIR ACTIVITY

DAILY ACTIVITY - VISITORS - INSPECTIONS

Supervisor's Signature: ______________________________

DAILY LOG

Day: ____________________

April 4, 20____ **Weather: 8:00 A.M. ___ Noon: ___ 4:00 P.M. ___**

WORK FORCE

Supt.	____
Foreman	____
Carpenters	____
Laborers	____
Masons	____
Oper, Engrs.	____
Iron Wrkers	____
Electricians	____
Plumbers	____
Steam Fittrs	____
Sheet Metal	____
Glazers	____
Roofers	____
Sprinkler	____
Painters	____
Tile Setters	____
Carpet Lyrs	____
Controls	____
________	____
________	____
________	____
________	____

SUBCONTRACTORS ON PROJECT & THEIR ACTIVITY

DAILY ACTIVITY - VISITORS - INSPECTIONS

Supervisor's Signature: ________________________________

DAILY LOG

Day: ____________________

April 5, 20____

Weather: 8:00 A.M. ___ Noon: ___ 4:00 P.M. ___

WORK FORCE

Supt.	____
Foreman	____
Carpenters	____
Laborers	____
Masons	____
Oper, Engrs.	____
Iron Wrkers	____
Electricians	____
Plumbers	____
Steam Fittrs	____
Sheet Metal	____
Glazers	____
Roofers	____
Sprinkler	____
Painters	____
Tile Setters	____
Carpet Lyrs	____
Controls	____
________	____
________	____
________	____
________	____

SUBCONTRACTORS ON PROJECT & THEIR ACTIVITY

DAILY ACTIVITY - VISITORS - INSPECTIONS

Supervisor's Signature: ______________________________

DAILY LOG

Day: ____________________

April 6, 20____ **Weather: 8:00 A.M. ____ Noon: ____ 4:00 P.M. ____**

WORK FORCE	
Supt.	____
Foreman	____
Carpenters	____
Laborers	____
Masons	____
Oper, Engrs.	____
Iron Wrkers	____
Electricians	____
Plumbers	____
Steam Fittrs	____
Sheet Metal	____
Glazers	____
Roofers	____
Sprinkler	____
Painters	____
Tile Setters	____
Carpet Lyrs	____
Controls	____
________	____
________	____
________	____
________	____

SUBCONTRACTORS ON PROJECT & THEIR ACTIVITY

DAILY ACTIVITY - VISITORS - INSPECTIONS

Supervisor's Signature: ______________________________

DAILY LOG

Day: ________________

April 7, 20____ **Weather: 8:00 A.M.** ____ **Noon:** ____ **4:00 P.M.** ____

WORK FORCE

Supt.	____
Foreman	____
Carpenters	____
Laborers	____
Masons	____
Oper, Engrs.	____
Iron Wrkers	____
Electricians	____
Plumbers	____
Steam Fittrs	____
Sheet Metal	____
Glazers	____
Roofers	____
Sprinkler	____
Painters	____
Tile Setters	____
Carpet Lyrs	____
Controls	____
________	____
________	____
________	____
________	____

SUBCONTRACTORS ON PROJECT & THEIR ACTIVITY

____

____

____

DAILY ACTIVITY - VISITORS - INSPECTIONS

____

____

____

Supervisor's Signature: ______________________________

DAILY LOG

Day: ____________________

April 8, 20____ **Weather: 8:00 A.M. ____ Noon: ____ 4:00 P.M. ____**

WORK FORCE

Supt.	____
Foreman	____
Carpenters	____
Laborers	____
Masons	____
Oper, Engrs.	____
Iron Wrkers	____
Electricians	____
Plumbers	____
Steam Fittrs	____
Sheet Metal	____
Glazers	____
Roofers	____
Sprinkler	____
Painters	____
Tile Setters	____
Carpet Lyrs	____
Controls	____
________	____
________	____
________	____
________	____

SUBCONTRACTORS ON PROJECT & THEIR ACTIVITY

DAILY ACTIVITY - VISITORS - INSPECTIONS

Supervisor's Signature: ______________________________

DAILY LOG

Day: ____________________

April 9, 20____

Weather: 8:00 A.M. ___ Noon: ___ 4:00 P.M. ___

WORK FORCE

Supt.	____
Foreman	____
Carpenters	____
Laborers	____
Masons	____
Oper, Engrs.	____
Iron Wrkers	____
Electricians	____
Plumbers	____
Steam Fittrs	____
Sheet Metal	____
Glazers	____
Roofers	____
Sprinkler	____
Painters	____
Tile Setters	____
Carpet Lyrs	____
Controls	____
________	____
________	____
________	____
________	____

SUBCONTRACTORS ON PROJECT & THEIR ACTIVITY

DAILY ACTIVITY - VISITORS - INSPECTIONS

Supervisor's Signature: ______________________________

DAILY LOG

Day: ______________

April 10, 20___ **Weather: 8:00 A.M. ___ Noon: ___ 4:00 P.M. ___**

WORK FORCE

Supt.	___
Foreman	___
Carpenters	___
Laborers	___
Masons	___
Oper, Engrs.	___
Iron Wrkers	___
Electricians	___
Plumbers	___
Steam Fittrs	___
Sheet Metal	___
Glazers	___
Roofers	___
Sprinkler	___
Painters	___
Tile Setters	___
Carpet Lyrs	___
Controls	___
___	___
___	___
___	___
___	___

SUBCONTRACTORS ON PROJECT & THEIR ACTIVITY

DAILY ACTIVITY - VISITORS - INSPECTIONS

Supervisor's Signature: ______________________________

DAILY LOG

Day: ____________________

April 11, 20____

Weather: 8:00 A.M. ____ Noon: ____ 4:00 P.M. ____

WORK FORCE

Supt.	____
Foreman	____
Carpenters	____
Laborers	____
Masons	____
Oper, Engrs.	____
Iron Wrkers	____
Electricians	____
Plumbers	____
Steam Fittrs	____
Sheet Metal	____
Glazers	____
Roofers	____
Sprinkler	____
Painters	____
Tile Setters	____
Carpet Lyrs	____
Controls	____
________	____
________	____
________	____
________	____

SUBCONTRACTORS ON PROJECT & THEIR ACTIVITY

DAILY ACTIVITY - VISITORS - INSPECTIONS

Supervisor's Signature: ______________________________

DAILY LOG

Day: ____________________

April 12, 20____ **Weather: 8:00 A.M. ____ Noon: ____ 4:00 P.M. ____**

WORK FORCE	
Supt.	____
Foreman	____
Carpenters	____
Laborers	____
Masons	____
Oper, Engrs.	____
Iron Wrkers	____
Electricians	____
Plumbers	____
Steam Fittrs	____
Sheet Metal	____
Glazers	____
Roofers	____
Sprinkler	____
Painters	____
Tile Setters	____
Carpet Lyrs	____
Controls	____
________	____
________	____
________	____
________	____

SUBCONTRACTORS ON PROJECT & THEIR ACTIVITY

DAILY ACTIVITY - VISITORS - INSPECTIONS

Supervisor's Signature: ______________________________

DAILY LOG

Day: ________________

April 13, 20____ **Weather: 8:00 A.M. ____ Noon: ____ 4:00 P.M. ____**

WORK FORCE	
Supt.	____
Foreman	____
Carpenters	____
Laborers	____
Masons	____
Oper, Engrs.	____
Iron Wrkers	____
Electricians	____
Plumbers	____
Steam Fittrs	____
Sheet Metal	____
Glazers	____
Roofers	____
Sprinkler	____
Painters	____
Tile Setters	____
Carpet Lyrs	____
Controls	____
________	____
________	____
________	____
________	____

SUBCONTRACTORS ON PROJECT & THEIR ACTIVITY

DAILY ACTIVITY - VISITORS - INSPECTIONS

Supervisor's Signature: ________________________________

DAILY LOG

Day: ____________________

April 14, 20____ **Weather: 8:00 A.M. ___ Noon: ___ 4:00 P.M. ___**

WORK FORCE

Supt.	___
Foreman	___
Carpenters	___
Laborers	___
Masons	___
Oper, Engrs.	___
Iron Wrkers	___
Electricians	___
Plumbers	___
Steam Fittrs	___
Sheet Metal	___
Glazers	___
Roofers	___
Sprinkler	___
Painters	___
Tile Setters	___
Carpet Lyrs	___
Controls	___
________	___
________	___
________	___
________	___

SUBCONTRACTORS ON PROJECT & THEIR ACTIVITY

DAILY ACTIVITY - VISITORS - INSPECTIONS

Supervisor's Signature: ____________________

DAILY LOG

Day: ____________________

April 15, 20____ **Weather: 8:00 A.M. ___ Noon: ___ 4:00 P.M. ___**

WORK FORCE

Supt.	___
Foreman	___
Carpenters	___
Laborers	___
Masons	___
Oper, Engrs.	___
Iron Wrkers	___
Electricians	___
Plumbers	___
Steam Fittrs	___
Sheet Metal	___
Glazers	___
Roofers	___
Sprinkler	___
Painters	___
Tile Setters	___
Carpet Lyrs	___
Controls	___
________	___
________	___
________	___
________	___

SUBCONTRACTORS ON PROJECT & THEIR ACTIVITY

DAILY ACTIVITY - VISITORS - INSPECTIONS

Supervisor's Signature: ______________________________

DAILY LOG

Day: ______________

April 16, 20____ **Weather: 8:00 A.M. ___ Noon: ___ 4:00 P.M. ___**

WORK FORCE	
Supt.	____
Foreman	____
Carpenters	____
Laborers	____
Masons	____
Oper, Engrs.	____
Iron Wrkers	____
Electricians	____
Plumbers	____
Steam Fittrs	____
Sheet Metal	____
Glazers	____
Roofers	____
Sprinkler	____
Painters	____
Tile Setters	____
Carpet Lyrs	____
Controls	____
________	____
________	____
________	____
________	____

SUBCONTRACTORS ON PROJECT & THEIR ACTIVITY

DAILY ACTIVITY - VISITORS - INSPECTIONS

Supervisor's Signature: ______________________________

DAILY LOG

Day: ______________

April 17, 20____ **Weather: 8:00 A.M. ___ Noon: ___ 4:00 P.M. ___**

WORK FORCE		SUBCONTRACTORS ON PROJECT & THEIR ACTIVITY
Supt.	___	
Foreman	___	
Carpenters	___	
Laborers	___	
Masons	___	
Oper, Engrs.	___	
Iron Wrkers	___	
Electricians	___	
Plumbers	___	
Steam Fittrs	___	
Sheet Metal	___	
Glazers	___	
Roofers	___	
Sprinkler	___	
Painters	___	
Tile Setters	___	
Carpet Lyrs	___	
Controls	___	
______	___	DAILY ACTIVITY - VISITORS - INSPECTIONS
______	___	
______	___	
______	___	

Supervisor's Signature: ______________________

DAILY LOG

Day: ______________

April 18, 20____

Weather: 8:00 A.M. ___ Noon: ___ 4:00 P.M. ___

WORK FORCE	
Supt.	___
Foreman	___
Carpenters	___
Laborers	___
Masons	___
Oper, Engrs.	___
Iron Wrkers	___
Electricians	___
Plumbers	___
Steam Fittrs	___
Sheet Metal	___
Glazers	___
Roofers	___
Sprinkler	___
Painters	___
Tile Setters	___
Carpet Lyrs	___
Controls	___
______	___
______	___
______	___
______	___

SUBCONTRACTORS ON PROJECT & THEIR ACTIVITY

DAILY ACTIVITY - VISITORS - INSPECTIONS

Supervisor's Signature: ______________________

DAILY LOG

Day: ____________________

April 19, 20____ **Weather: 8:00 A.M. ____ Noon: ____ 4:00 P.M. ____**

WORK FORCE

Supt.	____
Foreman	____
Carpenters	____
Laborers	____
Masons	____
Oper, Engrs.	____
Iron Wrkers	____
Electricians	____
Plumbers	____
Steam Fittrs	____
Sheet Metal	____
Glazers	____
Roofers	____
Sprinkler	____
Painters	____
Tile Setters	____
Carpet Lyrs	____
Controls	____
________	____
________	____
________	____
________	____

SUBCONTRACTORS ON PROJECT & THEIR ACTIVITY

DAILY ACTIVITY - VISITORS - INSPECTIONS

Supervisor's Signature: ____________________

DAILY LOG

Day: ________________

April 20, 20____ **Weather: 8:00 A.M. ___ Noon: ___ 4:00 P.M. ___**

WORK FORCE

Supt.	____
Foreman	____
Carpenters	____
Laborers	____
Masons	____
Oper, Engrs.	____
Iron Wrkers	____
Electricians	____
Plumbers	____
Steam Fittrs	____
Sheet Metal	____
Glazers	____
Roofers	____
Sprinkler	____
Painters	____
Tile Setters	____
Carpet Lyrs	____
Controls	____
________	____
________	____
________	____
________	____

SUBCONTRACTORS ON PROJECT & THEIR ACTIVITY

DAILY ACTIVITY - VISITORS - INSPECTIONS

Supervisor's Signature: ____________________

DAILY LOG

Day: ____________________

April 21, 20____ **Weather: 8:00 A.M. ___ Noon: ___ 4:00 P.M. ___**

WORK FORCE

Supt.	____
Foreman	____
Carpenters	____
Laborers	____
Masons	____
Oper, Engrs.	____
Iron Wrkers	____
Electricians	____
Plumbers	____
Steam Fittrs	____
Sheet Metal	____
Glazers	____
Roofers	____
Sprinkler	____
Painters	____
Tile Setters	____
Carpet Lyrs	____
Controls	____
________	____
________	____
________	____
________	____

SUBCONTRACTORS ON PROJECT & THEIR ACTIVITY

DAILY ACTIVITY - VISITORS - INSPECTIONS

Supervisor's Signature: ______________________________

DAILY LOG

Day: ______________________

April 22, 20____

Weather: 8:00 A.M. ____ Noon: ____ 4:00 P.M. ____

WORK FORCE	
Supt.	____
Foreman	____
Carpenters	____
Laborers	____
Masons	____
Oper, Engrs.	____
Iron Wrkers	____
Electricians	____
Plumbers	____
Steam Fittrs	____
Sheet Metal	____
Glazers	____
Roofers	____
Sprinkler	____
Painters	____
Tile Setters	____
Carpet Lyrs	____
Controls	____
________	____
________	____
________	____
________	____

SUBCONTRACTORS ON PROJECT & THEIR ACTIVITY

DAILY ACTIVITY - VISITORS - INSPECTIONS

Supervisor's Signature: ______________________

DAILY LOG

Day: ____________________

April 23, 20____ **Weather: 8:00 A.M. ____ Noon: ____ 4:00 P.M. ____**

WORK FORCE

Supt.	____
Foreman	____
Carpenters	____
Laborers	____
Masons	____
Oper, Engrs.	____
Iron Wrkers	____
Electricians	____
Plumbers	____
Steam Fittrs	____
Sheet Metal	____
Glazers	____
Roofers	____
Sprinkler	____
Painters	____
Tile Setters	____
Carpet Lyrs	____
Controls	____
________	____
________	____
________	____
________	____

SUBCONTRACTORS ON PROJECT & THEIR ACTIVITY

DAILY ACTIVITY - VISITORS - INSPECTIONS

Supervisor's Signature: ______________________________

DAILY LOG

Day: ____________________

April 24, 20____ **Weather: 8:00 A.M. ___ Noon: ___ 4:00 P.M. ___**

WORK FORCE

Supt.	____
Foreman	____
Carpenters	____
Laborers	____
Masons	____
Oper, Engrs.	____
Iron Wrkers	____
Electricians	____
Plumbers	____
Steam Fittrs	____
Sheet Metal	____
Glazers	____
Roofers	____
Sprinkler	____
Painters	____
Tile Setters	____
Carpet Lyrs	____
Controls	____
________	____
________	____
________	____
________	____

SUBCONTRACTORS ON PROJECT & THEIR ACTIVITY

DAILY ACTIVITY - VISITORS - INSPECTIONS

Supervisor's Signature: ______________________________

DAILY LOG

Day: ________________

April 25, 20____

Weather: 8:00 A.M. ___ Noon: ___ 4:00 P.M. ___

WORK FORCE	
Supt.	____
Foreman	____
Carpenters	____
Laborers	____
Masons	____
Oper, Engrs.	____
Iron Wrkers	____
Electricians	____
Plumbers	____
Steam Fittrs	____
Sheet Metal	____
Glazers	____
Roofers	____
Sprinkler	____
Painters	____
Tile Setters	____
Carpet Lyrs	____
Controls	____
________	____
________	____
________	____
________	____

SUBCONTRACTORS ON PROJECT & THEIR ACTIVITY

DAILY ACTIVITY - VISITORS - INSPECTIONS

Supervisor's Signature: ________________________

DAILY LOG

Day: ________________

April 26, 20___ **Weather: 8:00 A.M. ___ Noon: ___ 4:00 P.M. ___**

WORK FORCE	
Supt.	___
Foreman	___
Carpenters	___
Laborers	___
Masons	___
Oper, Engrs.	___
Iron Wrkers	___
Electricians	___
Plumbers	___
Steam Fittrs	___
Sheet Metal	___
Glazers	___
Roofers	___
Sprinkler	___
Painters	___
Tile Setters	___
Carpet Lyrs	___
Controls	___
______	___
______	___
______	___
______	___

SUBCONTRACTORS ON PROJECT & THEIR ACTIVITY

DAILY ACTIVITY - VISITORS - INSPECTIONS

Supervisor's Signature: ______________________________

DAILY LOG

Day: ____________________

April 27, 20____ **Weather: 8:00 A.M. ___ Noon: ___ 4:00 P.M. ___**

WORK FORCE	
Supt.	___
Foreman	___
Carpenters	___
Laborers	___
Masons	___
Oper, Engrs.	___
Iron Wrkers	___
Electricians	___
Plumbers	___
Steam Fittrs	___
Sheet Metal	___
Glazers	___
Roofers	___
Sprinkler	___
Painters	___
Tile Setters	___
Carpet Lyrs	___
Controls	___
________	___
________	___
________	___
________	___

SUBCONTRACTORS ON PROJECT & THEIR ACTIVITY

DAILY ACTIVITY - VISITORS - INSPECTIONS

Supervisor's Signature: ______________________________

DAILY LOG

Day: ____________________

April 28, 20____ **Weather: 8:00 A.M. ___ Noon: ___ 4:00 P.M. ___**

WORK FORCE	
Supt.	___
Foreman	___
Carpenters	___
Laborers	___
Masons	___
Oper, Engrs.	___
Iron Wrkers	___
Electricians	___
Plumbers	___
Steam Fittrs	___
Sheet Metal	___
Glazers	___
Roofers	___
Sprinkler	___
Painters	___
Tile Setters	___
Carpet Lyrs	___
Controls	___
________	___
________	___
________	___
________	___

SUBCONTRACTORS ON PROJECT & THEIR ACTIVITY

DAILY ACTIVITY - VISITORS - INSPECTIONS

Supervisor's Signature: ______________________________

DAILY LOG

Day: ____________________

April 29, 20____

Weather: 8:00 A.M. ____ Noon: ____ 4:00 P.M. ____

WORK FORCE

Supt.	____
Foreman	____
Carpenters	____
Laborers	____
Masons	____
Oper, Engrs.	____
Iron Wrkers	____
Electricians	____
Plumbers	____
Steam Fittrs	____
Sheet Metal	____
Glazers	____
Roofers	____
Sprinkler	____
Painters	____
Tile Setters	____
Carpet Lyrs	____
Controls	____
________	____
________	____
________	____
________	____

SUBCONTRACTORS ON PROJECT & THEIR ACTIVITY

DAILY ACTIVITY - VISITORS - INSPECTIONS

Supervisor's Signature: ______________________________

DAILY LOG

Day: ____________________

April 30, 20____ **Weather: 8:00 A.M.** ____ **Noon:** ____ **4:00 P.M.** ____

WORK FORCE

Supt.	____
Foreman	____
Carpenters	____
Laborers	____
Masons	____
Oper, Engrs.	____
Iron Wrkers	____
Electricians	____
Plumbers	____
Steam Fittrs	____
Sheet Metal	____
Glazers	____
Roofers	____
Sprinkler	____
Painters	____
Tile Setters	____
Carpet Lyrs	____
Controls	____
____	____
____	____
____	____
____	____

SUBCONTRACTORS ON PROJECT & THEIR ACTIVITY

DAILY ACTIVITY - VISITORS - INSPECTIONS

Supervisor's Signature: ______________________________

DAILY LOG

Day: ____________________

May 1, 20____ **Weather: 8:00 A.M. ____ Noon: ____ 4:00 P.M. ____**

WORK FORCE

Supt.	____
Foreman	____
Carpenters	____
Laborers	____
Masons	____
Oper, Engrs.	____
Iron Wrkers	____
Electricians	____
Plumbers	____
Steam Fittrs	____
Sheet Metal	____
Glazers	____
Roofers	____
Sprinkler	____
Painters	____
Tile Setters	____
Carpet Lyrs	____
Controls	____
________	____
________	____
________	____
________	____

SUBCONTRACTORS ON PROJECT & THEIR ACTIVITY

DAILY ACTIVITY - VISITORS - INSPECTIONS

Supervisor's Signature: ______________________________

DAILY LOG

Day: ____________________

May 2, 20____

Weather: 8:00 A.M. ___ Noon: ___ 4:00 P.M. ___

WORK FORCE	
Supt.	____
Foreman	____
Carpenters	____
Laborers	____
Masons	____
Oper, Engrs.	____
Iron Wrkers	____
Electricians	____
Plumbers	____
Steam Fittrs	____
Sheet Metal	____
Glazers	____
Roofers	____
Sprinkler	____
Painters	____
Tile Setters	____
Carpet Lyrs	____
Controls	____
________	____
________	____
________	____
________	____

SUBCONTRACTORS ON PROJECT & THEIR ACTIVITY

DAILY ACTIVITY - VISITORS - INSPECTIONS

Supervisor's Signature: ____________________

DAILY LOG

Day: ____________________

May 3, 20____ **Weather: 8:00 A.M.** ____ **Noon:** ____ **4:00 P.M.** ____

WORK FORCE

Supt.	____
Foreman	____
Carpenters	____
Laborers	____
Masons	____
Oper, Engrs.	____
Iron Wrkers	____
Electricians	____
Plumbers	____
Steam Fittrs	____
Sheet Metal	____
Glazers	____
Roofers	____
Sprinkler	____
Painters	____
Tile Setters	____
Carpet Lyrs	____
Controls	____
________	____
________	____
________	____
________	____

SUBCONTRACTORS ON PROJECT & THEIR ACTIVITY

DAILY ACTIVITY - VISITORS - INSPECTIONS

Supervisor's Signature: ______________________________

DAILY LOG

Day: ________________

May 4, 20____ **Weather: 8:00 A.M. ___ Noon: ___ 4:00 P.M. ___**

WORK FORCE		SUBCONTRACTORS ON PROJECT & THEIR ACTIVITY
Supt.	___	
Foreman	___	
Carpenters	___	
Laborers	___	
Masons	___	
Oper, Engrs.	___	
Iron Wrkers	___	
Electricians	___	
Plumbers	___	
Steam Fittrs	___	
Sheet Metal	___	
Glazers	___	
Roofers	___	
Sprinkler	___	
Painters	___	
Tile Setters	___	
Carpet Lyrs	___	
Controls	___	
___	___	DAILY ACTIVITY - VISITORS - INSPECTIONS
___	___	
___	___	
___	___	

Supervisor's Signature: ________________

DAILY LOG

Day: ____________________

May 5, 20___

Weather: 8:00 A.M. ___ Noon: ___ 4:00 P.M. ___

WORK FORCE

Supt.	___
Foreman	___
Carpenters	___
Laborers	___
Masons	___
Oper, Engrs.	___
Iron Wrkers	___
Electricians	___
Plumbers	___
Steam Fittrs	___
Sheet Metal	___
Glazers	___
Roofers	___
Sprinkler	___
Painters	___
Tile Setters	___
Carpet Lyrs	___
Controls	___
___	___
___	___
___	___
___	___

SUBCONTRACTORS ON PROJECT & THEIR ACTIVITY

DAILY ACTIVITY - VISITORS - INSPECTIONS

Supervisor's Signature: ____________________

DAILY LOG

Day: ____________________

May 6, 20____ **Weather: 8:00 A.M. ___ Noon: ___ 4:00 P.M. ___**

WORK FORCE	
Supt.	____
Foreman	____
Carpenters	____
Laborers	____
Masons	____
Oper, Engrs.	____
Iron Wrkers	____
Electricians	____
Plumbers	____
Steam Fittrs	____
Sheet Metal	____
Glazers	____
Roofers	____
Sprinkler	____
Painters	____
Tile Setters	____
Carpet Lyrs	____
Controls	____
________	____
________	____
________	____
________	____

SUBCONTRACTORS ON PROJECT & THEIR ACTIVITY

DAILY ACTIVITY - VISITORS - INSPECTIONS

Supervisor's Signature: ______________________________

DAILY LOG

Day: ________________

May 7, 20___ **Weather: 8:00 A.M. ___ Noon: ___ 4:00 P.M. ___**

WORK FORCE

Supt.	____
Foreman	____
Carpenters	____
Laborers	____
Masons	____
Oper, Engrs.	____
Iron Wrkers	____
Electricians	____
Plumbers	____
Steam Fittrs	____
Sheet Metal	____
Glazers	____
Roofers	____
Sprinkler	____
Painters	____
Tile Setters	____
Carpet Lyrs	____
Controls	____
________	____
________	____
________	____
________	____

SUBCONTRACTORS ON PROJECT & THEIR ACTIVITY

DAILY ACTIVITY - VISITORS - INSPECTIONS

Supervisor's Signature: ______________________________

DAILY LOG

Day: ________________

May 8, 20___ **Weather: 8:00 A.M. ___ Noon: ___ 4:00 P.M. ___**

WORK FORCE

Supt.	___
Foreman	___
Carpenters	___
Laborers	___
Masons	___
Oper, Engrs.	___
Iron Wrkers	___
Electricians	___
Plumbers	___
Steam Fittrs	___
Sheet Metal	___
Glazers	___
Roofers	___
Sprinkler	___
Painters	___
Tile Setters	___
Carpet Lyrs	___
Controls	___
______	___
______	___
______	___
______	___

SUBCONTRACTORS ON PROJECT & THEIR ACTIVITY

DAILY ACTIVITY - VISITORS - INSPECTIONS

Supervisor's Signature: ______________________

DAILY LOG

Day: ______________________

May 9, 20____

Weather: 8:00 A.M. ___ Noon: ___ 4:00 P.M. ___

WORK FORCE	
Supt.	____
Foreman	____
Carpenters	____
Laborers	____
Masons	____
Oper, Engrs.	____
Iron Wrkers	____
Electricians	____
Plumbers	____
Steam Fittrs	____
Sheet Metal	____
Glazers	____
Roofers	____
Sprinkler	____
Painters	____
Tile Setters	____
Carpet Lyrs	____
Controls	____
________	____
________	____
________	____
________	____

SUBCONTRACTORS ON PROJECT & THEIR ACTIVITY

DAILY ACTIVITY - VISITORS - INSPECTIONS

Supervisor's Signature: ______________________________

DAILY LOG

Day: ____________________

May 10, 20____ **Weather: 8:00 A.M. ____ Noon: ____ 4:00 P.M. ____**

WORK FORCE		**SUBCONTRACTORS ON PROJECT & THEIR ACTIVITY**
Supt.	____	____________________
Foreman	____	____________________
Carpenters	____	____________________
Laborers	____	____________________
Masons	____	____________________
Oper, Engrs.	____	____________________
Iron Wrkers	____	____________________
Electricians	____	____________________
Plumbers	____	____________________
Steam Fittrs	____	____________________
Sheet Metal	____	____________________
Glazers	____	____________________
Roofers	____	____________________
Sprinkler	____	____________________
Painters	____	____________________
Tile Setters	____	____________________
Carpet Lyrs	____	____________________
Controls	____	____________________
________	____	DAILY ACTIVITY - VISITORS - INSPECTIONS
________	____	____________________
________	____	____________________
________	____	____________________
		____________________

____

____

____

Supervisor's Signature: ______________________________

DAILY LOG

Day: ____________________

May 11, 20____ **Weather: 8:00 A.M. ____ Noon: ____ 4:00 P.M. ____**

WORK FORCE	
Supt.	____
Foreman	____
Carpenters	____
Laborers	____
Masons	____
Oper, Engrs.	____
Iron Wrkers	____
Electricians	____
Plumbers	____
Steam Fittrs	____
Sheet Metal	____
Glazers	____
Roofers	____
Sprinkler	____
Painters	____
Tile Setters	____
Carpet Lyrs	____
Controls	____
________	____
________	____
________	____
________	____

SUBCONTRACTORS ON PROJECT & THEIR ACTIVITY

DAILY ACTIVITY - VISITORS - INSPECTIONS

Supervisor's Signature: ______________________________

DAILY LOG

Day: ____________________

May 12, 20____ **Weather: 8:00 A.M. ____ Noon: ____ 4:00 P.M. ____**

WORK FORCE	
Supt.	____
Foreman	____
Carpenters	____
Laborers	____
Masons	____
Oper, Engrs.	____
Iron Wrkers	____
Electricians	____
Plumbers	____
Steam Fittrs	____
Sheet Metal	____
Glazers	____
Roofers	____
Sprinkler	____
Painters	____
Tile Setters	____
Carpet Lyrs	____
Controls	____
________	____
________	____
________	____
________	____

SUBCONTRACTORS ON PROJECT & THEIR ACTIVITY

DAILY ACTIVITY - VISITORS - INSPECTIONS

Supervisor's Signature: ______________________________

DAILY LOG

Day: ______________

May 13, 20___ **Weather: 8:00 A.M. ___ Noon: ___ 4:00 P.M. ___**

WORK FORCE

Supt.	___
Foreman	___
Carpenters	___
Laborers	___
Masons	___
Oper, Engrs.	___
Iron Wrkers	___
Electricians	___
Plumbers	___
Steam Fittrs	___
Sheet Metal	___
Glazers	___
Roofers	___
Sprinkler	___
Painters	___
Tile Setters	___
Carpet Lyrs	___
Controls	___
________	___
________	___
________	___
________	___

SUBCONTRACTORS ON PROJECT & THEIR ACTIVITY

DAILY ACTIVITY - VISITORS - INSPECTIONS

Supervisor's Signature: ______________________

DAILY LOG

Day: ____________________

May 14, 20____ **Weather: 8:00 A.M. ___ Noon: ___ 4:00 P.M. ___**

WORK FORCE	
Supt.	___
Foreman	___
Carpenters	___
Laborers	___
Masons	___
Oper, Engrs.	___
Iron Wrkers	___
Electricians	___
Plumbers	___
Steam Fittrs	___
Sheet Metal	___
Glazers	___
Roofers	___
Sprinkler	___
Painters	___
Tile Setters	___
Carpet Lyrs	___
Controls	___
___	___
___	___
___	___
___	___

SUBCONTRACTORS ON PROJECT & THEIR ACTIVITY

DAILY ACTIVITY - VISITORS - INSPECTIONS

Supervisor's Signature: ____________________

DAILY LOG

Day: ______________

May 15, 20___

Weather: 8:00 A.M. ___ Noon: ___ 4:00 P.M. ___

WORK FORCE

Supt.	___
Foreman	___
Carpenters	___
Laborers	___
Masons	___
Oper, Engrs.	___
Iron Wrkers	___
Electricians	___
Plumbers	___
Steam Fittrs	___
Sheet Metal	___
Glazers	___
Roofers	___
Sprinkler	___
Painters	___
Tile Setters	___
Carpet Lyrs	___
Controls	___
______	___
______	___
______	___
______	___

SUBCONTRACTORS ON PROJECT & THEIR ACTIVITY

DAILY ACTIVITY - VISITORS - INSPECTIONS

Supervisor's Signature: ______________________

DAILY LOG

Day: ________________

May 16, 20___ **Weather: 8:00 A.M. ___ Noon: ___ 4:00 P.M. ___**

WORK FORCE	
Supt.	___
Foreman	___
Carpenters	___
Laborers	___
Masons	___
Oper, Engrs.	___
Iron Wrkers	___
Electricians	___
Plumbers	___
Steam Fittrs	___
Sheet Metal	___
Glazers	___
Roofers	___
Sprinkler	___
Painters	___
Tile Setters	___
Carpet Lyrs	___
Controls	___
______	___
______	___
______	___
______	___

SUBCONTRACTORS ON PROJECT & THEIR ACTIVITY

DAILY ACTIVITY - VISITORS - INSPECTIONS

Supervisor's Signature: ________________________________

DAILY LOG

Day: ______________

May 17, 20___ **Weather: 8:00 A.M. ___ Noon: ___ 4:00 P.M. ___**

WORK FORCE	
Supt.	___
Foreman	___
Carpenters	___
Laborers	___
Masons	___
Oper, Engrs.	___
Iron Wrkers	___
Electricians	___
Plumbers	___
Steam Fittrs	___
Sheet Metal	___
Glazers	___
Roofers	___
Sprinkler	___
Painters	___
Tile Setters	___
Carpet Lyrs	___
Controls	___
______	___
______	___
______	___
______	___

SUBCONTRACTORS ON PROJECT & THEIR ACTIVITY

DAILY ACTIVITY - VISITORS - INSPECTIONS

Supervisor's Signature: ______________

DAILY LOG

Day: ____________________

May 18, 20____ **Weather: 8:00 A.M. ___ Noon: ___ 4:00 P.M. ___**

WORK FORCE	
Supt.	___
Foreman	___
Carpenters	___
Laborers	___
Masons	___
Oper, Engrs.	___
Iron Wrkers	___
Electricians	___
Plumbers	___
Steam Fittrs	___
Sheet Metal	___
Glazers	___
Roofers	___
Sprinkler	___
Painters	___
Tile Setters	___
Carpet Lyrs	___
Controls	___
________	___
________	___
________	___
________	___

SUBCONTRACTORS ON PROJECT & THEIR ACTIVITY

DAILY ACTIVITY - VISITORS - INSPECTIONS

Supervisor's Signature: ______________________________

DAILY LOG

Day: ____________________

May 19, 20____ **Weather: 8:00 A.M. ___ Noon: ___ 4:00 P.M. ___**

WORK FORCE

Supt.	____
Foreman	____
Carpenters	____
Laborers	____
Masons	____
Oper, Engrs.	____
Iron Wrkers	____
Electricians	____
Plumbers	____
Steam Fittrs	____
Sheet Metal	____
Glazers	____
Roofers	____
Sprinkler	____
Painters	____
Tile Setters	____
Carpet Lyrs	____
Controls	____
________	____
________	____
________	____
________	____

SUBCONTRACTORS ON PROJECT & THEIR ACTIVITY

DAILY ACTIVITY - VISITORS - INSPECTIONS

Supervisor's Signature: ______________________________

DAILY LOG

Day: ____________________

May 20, 20____ **Weather: 8:00 A.M.** ___ **Noon:** ___ **4:00 P.M.** ___

WORK FORCE

Supt.	____
Foreman	____
Carpenters	____
Laborers	____
Masons	____
Oper, Engrs.	____
Iron Wrkers	____
Electricians	____
Plumbers	____
Steam Fittrs	____
Sheet Metal	____
Glazers	____
Roofers	____
Sprinkler	____
Painters	____
Tile Setters	____
Carpet Lyrs	____
Controls	____
________	____
________	____
________	____
________	____

SUBCONTRACTORS ON PROJECT & THEIR ACTIVITY

DAILY ACTIVITY - VISITORS - INSPECTIONS

Supervisor's Signature: ______________________________

DAILY LOG

Day: ____________________

May 21, 20____ **Weather: 8:00 A.M. ____ Noon: ____ 4:00 P.M. ____**

WORK FORCE

Supt.	____
Foreman	____
Carpenters	____
Laborers	____
Masons	____
Oper, Engrs.	____
Iron Wrkers	____
Electricians	____
Plumbers	____
Steam Fittrs	____
Sheet Metal	____
Glazers	____
Roofers	____
Sprinkler	____
Painters	____
Tile Setters	____
Carpet Lyrs	____
Controls	____
________	____
________	____
________	____
________	____

SUBCONTRACTORS ON PROJECT & THEIR ACTIVITY

DAILY ACTIVITY - VISITORS - INSPECTIONS

Supervisor's Signature: ______________________________

DAILY LOG

Day: ____________________

May 22, 20____ **Weather: 8:00 A.M. ____ Noon: ____ 4:00 P.M. ____**

WORK FORCE

Supt.	____
Foreman	____
Carpenters	____
Laborers	____
Masons	____
Oper, Engrs.	____
Iron Wrkers	____
Electricians	____
Plumbers	____
Steam Fittrs	____
Sheet Metal	____
Glazers	____
Roofers	____
Sprinkler	____
Painters	____
Tile Setters	____
Carpet Lyrs	____
Controls	____
________	____
________	____
________	____
________	____

SUBCONTRACTORS ON PROJECT & THEIR ACTIVITY

DAILY ACTIVITY - VISITORS - INSPECTIONS

Supervisor's Signature: ______________________________

DAILY LOG

Day: ________________

May 23, 20___ **Weather: 8:00 A.M. ___ Noon: ___ 4:00 P.M. ___**

WORK FORCE

Supt.	___
Foreman	___
Carpenters	___
Laborers	___
Masons	___
Oper, Engrs.	___
Iron Wrkers	___
Electricians	___
Plumbers	___
Steam Fittrs	___
Sheet Metal	___
Glazers	___
Roofers	___
Sprinkler	___
Painters	___
Tile Setters	___
Carpet Lyrs	___
Controls	___
______	___
______	___
______	___
______	___

SUBCONTRACTORS ON PROJECT & THEIR ACTIVITY

DAILY ACTIVITY - VISITORS - INSPECTIONS

Supervisor's Signature: ______________________

DAILY LOG

Day: ____________________

May 24, 20____ **Weather: 8:00 A.M. ___ Noon: ___ 4:00 P.M. ___**

WORK FORCE	
Supt.	____
Foreman	____
Carpenters	____
Laborers	____
Masons	____
Oper, Engrs.	____
Iron Wrkers	____
Electricians	____
Plumbers	____
Steam Fittrs	____
Sheet Metal	____
Glazers	____
Roofers	____
Sprinkler	____
Painters	____
Tile Setters	____
Carpet Lyrs	____
Controls	____
________	____
________	____
________	____
________	____

SUBCONTRACTORS ON PROJECT & THEIR ACTIVITY

DAILY ACTIVITY - VISITORS - INSPECTIONS

Supervisor's Signature: ______________________________

DAILY LOG

Day: ____________________

May 25, 20____

Weather: 8:00 A.M. ___ Noon: ___ 4:00 P.M. ___

WORK FORCE

Supt.	___
Foreman	___
Carpenters	___
Laborers	___
Masons	___
Oper, Engrs.	___
Iron Wrkers	___
Electricians	___
Plumbers	___
Steam Fittrs	___
Sheet Metal	___
Glazers	___
Roofers	___
Sprinkler	___
Painters	___
Tile Setters	___
Carpet Lyrs	___
Controls	___
________	___
________	___
________	___
________	___

SUBCONTRACTORS ON PROJECT & THEIR ACTIVITY

DAILY ACTIVITY - VISITORS - INSPECTIONS

Supervisor's Signature: ________________________________

DAILY LOG

Day: ________________

May 26, 20___ **Weather: 8:00 A.M. ___ Noon: ___ 4:00 P.M. ___**

WORK FORCE	
Supt.	___
Foreman	___
Carpenters	___
Laborers	___
Masons	___
Oper, Engrs.	___
Iron Wrkers	___
Electricians	___
Plumbers	___
Steam Fittrs	___
Sheet Metal	___
Glazers	___
Roofers	___
Sprinkler	___
Painters	___
Tile Setters	___
Carpet Lyrs	___
Controls	___
___	___
___	___
___	___
___	___

SUBCONTRACTORS ON PROJECT & THEIR ACTIVITY

DAILY ACTIVITY - VISITORS - INSPECTIONS

Supervisor's Signature: ________________

DAILY LOG

Day: ____________________

May 27, 20____

Weather: 8:00 A.M. ____ Noon: ____ 4:00 P.M. ____

WORK FORCE	
Supt.	____
Foreman	____
Carpenters	____
Laborers	____
Masons	____
Oper, Engrs.	____
Iron Wrkers	____
Electricians	____
Plumbers	____
Steam Fittrs	____
Sheet Metal	____
Glazers	____
Roofers	____
Sprinkler	____
Painters	____
Tile Setters	____
Carpet Lyrs	____
Controls	____
________	____
________	____
________	____
________	____

SUBCONTRACTORS ON PROJECT & THEIR ACTIVITY

DAILY ACTIVITY - VISITORS - INSPECTIONS

Supervisor's Signature: ______________________________

DAILY LOG

Day: ________________

May 28, 20___ **Weather: 8:00 A.M. ___ Noon: ___ 4:00 P.M. ___**

WORK FORCE

Supt.	___
Foreman	___
Carpenters	___
Laborers	___
Masons	___
Oper, Engrs.	___
Iron Wrkers	___
Electricians	___
Plumbers	___
Steam Fittrs	___
Sheet Metal	___
Glazers	___
Roofers	___
Sprinkler	___
Painters	___
Tile Setters	___
Carpet Lyrs	___
Controls	___
________	___
________	___
________	___
________	___

SUBCONTRACTORS ON PROJECT & THEIR ACTIVITY

DAILY ACTIVITY - VISITORS - INSPECTIONS

Supervisor's Signature: ________________________________

DAILY LOG

Day: ________________

May 29, 20___ **Weather: 8:00 A.M.** ___ **Noon:** ___ **4:00 P.M.** ___

WORK FORCE	
Supt.	___
Foreman	___
Carpenters	___
Laborers	___
Masons	___
Oper, Engrs.	___
Iron Wrkers	___
Electricians	___
Plumbers	___
Steam Fittrs	___
Sheet Metal	___
Glazers	___
Roofers	___
Sprinkler	___
Painters	___
Tile Setters	___
Carpet Lyrs	___
Controls	___
________	___
________	___
________	___
________	___

SUBCONTRACTORS ON PROJECT & THEIR ACTIVITY

DAILY ACTIVITY - VISITORS - INSPECTIONS

Supervisor's Signature: ________________________________

DAILY LOG

Day: ____________________

May 30, 20____   **Weather: 8:00 A.M. ___ Noon: ___ 4:00 P.M. ___**

WORK FORCE

Supt.	___
Foreman	___
Carpenters	___
Laborers	___
Masons	___
Oper, Engrs.	___
Iron Wrkers	___
Electricians	___
Plumbers	___
Steam Fittrs	___
Sheet Metal	___
Glazers	___
Roofers	___
Sprinkler	___
Painters	___
Tile Setters	___
Carpet Lyrs	___
Controls	___
________	___
________	___
________	___
________	___

SUBCONTRACTORS ON PROJECT & THEIR ACTIVITY

DAILY ACTIVITY - VISITORS - INSPECTIONS

Supervisor's Signature: ______________________________

DAILY LOG

Day: ____________________

May 31, 20____

Weather: 8:00 A.M. ____ Noon: ____ 4:00 P.M. ____

WORK FORCE

Supt.	____
Foreman	____
Carpenters	____
Laborers	____
Masons	____
Oper, Engrs.	____
Iron Wrkers	____
Electricians	____
Plumbers	____
Steam Fittrs	____
Sheet Metal	____
Glazers	____
Roofers	____
Sprinkler	____
Painters	____
Tile Setters	____
Carpet Lyrs	____
Controls	____
________	____
________	____
________	____
________	____

SUBCONTRACTORS ON PROJECT & THEIR ACTIVITY

DAILY ACTIVITY - VISITORS - INSPECTIONS

Supervisor's Signature: ________________________________

DAILY LOG

Day: ________________

June 1, 20____ **Weather: 8:00 A.M. ___ Noon: ___ 4:00 P.M. ___**

WORK FORCE

Supt.	____
Foreman	____
Carpenters	____
Laborers	____
Masons	____
Oper, Engrs.	____
Iron Wrkers	____
Electricians	____
Plumbers	____
Steam Fittrs	____
Sheet Metal	____
Glazers	____
Roofers	____
Sprinkler	____
Painters	____
Tile Setters	____
Carpet Lyrs	____
Controls	____
________	____
________	____
________	____
________	____

SUBCONTRACTORS ON PROJECT & THEIR ACTIVITY

DAILY ACTIVITY - VISITORS - INSPECTIONS

Supervisor's Signature: ______________________________

DAILY LOG

Day: ____________________

June 2, 20____ **Weather: 8:00 A.M. ____ Noon: ____ 4:00 P.M. ____**

WORK FORCE

Supt.	____
Foreman	____
Carpenters	____
Laborers	____
Masons	____
Oper, Engrs.	____
Iron Wrkers	____
Electricians	____
Plumbers	____
Steam Fittrs	____
Sheet Metal	____
Glazers	____
Roofers	____
Sprinkler	____
Painters	____
Tile Setters	____
Carpet Lyrs	____
Controls	____
________	____
________	____
________	____
________	____

SUBCONTRACTORS ON PROJECT & THEIR ACTIVITY

DAILY ACTIVITY - VISITORS - INSPECTIONS

Supervisor's Signature: ______________________________

DAILY LOG

Day: ________________

June 3, 20___

Weather: 8:00 A.M. ___ Noon: ___ 4:00 P.M. ___

WORK FORCE

Supt.	___
Foreman	___
Carpenters	___
Laborers	___
Masons	___
Oper, Engrs.	___
Iron Wrkers	___
Electricians	___
Plumbers	___
Steam Fittrs	___
Sheet Metal	___
Glazers	___
Roofers	___
Sprinkler	___
Painters	___
Tile Setters	___
Carpet Lyrs	___
Controls	___
___	___
___	___
___	___
___	___

SUBCONTRACTORS ON PROJECT & THEIR ACTIVITY

DAILY ACTIVITY - VISITORS - INSPECTIONS

Supervisor's Signature: ______________________________

DAILY LOG

Day: ____________________

June 4, 20____

Weather: 8:00 A.M. ____ Noon: ____ 4:00 P.M. ____

WORK FORCE	
Supt.	____
Foreman	____
Carpenters	____
Laborers	____
Masons	____
Oper, Engrs.	____
Iron Wrkers	____
Electricians	____
Plumbers	____
Steam Fittrs	____
Sheet Metal	____
Glazers	____
Roofers	____
Sprinkler	____
Painters	____
Tile Setters	____
Carpet Lyrs	____
Controls	____
________	____
________	____
________	____
________	____

SUBCONTRACTORS ON PROJECT & THEIR ACTIVITY

DAILY ACTIVITY - VISITORS - INSPECTIONS

Supervisor's Signature: ______________________________

DAILY LOG

Day: ____________________

June 5, 20____

Weather: 8:00 A.M. ___ Noon: ___ 4:00 P.M. ___

WORK FORCE	
Supt.	___
Foreman	___
Carpenters	___
Laborers	___
Masons	___
Oper, Engrs.	___
Iron Wrkers	___
Electricians	___
Plumbers	___
Steam Fittrs	___
Sheet Metal	___
Glazers	___
Roofers	___
Sprinkler	___
Painters	___
Tile Setters	___
Carpet Lyrs	___
Controls	___
________	___
________	___
________	___
________	___

SUBCONTRACTORS ON PROJECT & THEIR ACTIVITY

DAILY ACTIVITY - VISITORS - INSPECTIONS

Supervisor's Signature: ______________________________

DAILY LOG

Day: ____________________

June 6, 20____

Weather: 8:00 A.M. ___ Noon: ___ 4:00 P.M. ___

WORK FORCE	
Supt.	____
Foreman	____
Carpenters	____
Laborers	____
Masons	____
Oper, Engrs.	____
Iron Wrkers	____
Electricians	____
Plumbers	____
Steam Fittrs	____
Sheet Metal	____
Glazers	____
Roofers	____
Sprinkler	____
Painters	____
Tile Setters	____
Carpet Lyrs	____
Controls	____
________	____
________	____
________	____
________	____

SUBCONTRACTORS ON PROJECT & THEIR ACTIVITY

DAILY ACTIVITY - VISITORS - INSPECTIONS

Supervisor's Signature: ______________________________

DAILY LOG

Day: ____________________

June 7, 20____ **Weather: 8:00 A.M. ___ Noon: ___ 4:00 P.M. ___**

WORK FORCE

Supt.	___
Foreman	___
Carpenters	___
Laborers	___
Masons	___
Oper, Engrs.	___
Iron Wrkers	___
Electricians	___
Plumbers	___
Steam Fittrs	___
Sheet Metal	___
Glazers	___
Roofers	___
Sprinkler	___
Painters	___
Tile Setters	___
Carpet Lyrs	___
Controls	___
______	___
______	___
______	___
______	___

SUBCONTRACTORS ON PROJECT & THEIR ACTIVITY

DAILY ACTIVITY - VISITORS - INSPECTIONS

Supervisor's Signature: ______________________________

DAILY LOG

Day: ____________________

June 8, 20____ **Weather: 8:00 A.M. ___ Noon: ___ 4:00 P.M. ___**

WORK FORCE

Supt.	___
Foreman	___
Carpenters	___
Laborers	___
Masons	___
Oper, Engrs.	___
Iron Wrkers	___
Electricians	___
Plumbers	___
Steam Fittrs	___
Sheet Metal	___
Glazers	___
Roofers	___
Sprinkler	___
Painters	___
Tile Setters	___
Carpet Lyrs	___
Controls	___
________	___
________	___
________	___
________	___

SUBCONTRACTORS ON PROJECT & THEIR ACTIVITY

DAILY ACTIVITY - VISITORS - INSPECTIONS

Supervisor's Signature: ____________________

DAILY LOG

Day: ______________________

June 9, 20____ **Weather: 8:00 A.M. ___ Noon: ___ 4:00 P.M. ___**

WORK FORCE	
Supt.	____
Foreman	____
Carpenters	____
Laborers	____
Masons	____
Oper, Engrs.	____
Iron Wrkers	____
Electricians	____
Plumbers	____
Steam Fittrs	____
Sheet Metal	____
Glazers	____
Roofers	____
Sprinkler	____
Painters	____
Tile Setters	____
Carpet Lyrs	____
Controls	____
________	____
________	____
________	____
________	____

SUBCONTRACTORS ON PROJECT & THEIR ACTIVITY

DAILY ACTIVITY - VISITORS - INSPECTIONS

Supervisor's Signature: ______________________________

DAILY LOG

Day: ______________

June 10, 20___ **Weather: 8:00 A.M.** ___ **Noon:** ___ **4:00 P.M.** ___

WORK FORCE

Supt.	___
Foreman	___
Carpenters	___
Laborers	___
Masons	___
Oper, Engrs.	___
Iron Wrkers	___
Electricians	___
Plumbers	___
Steam Fittrs	___
Sheet Metal	___
Glazers	___
Roofers	___
Sprinkler	___
Painters	___
Tile Setters	___
Carpet Lyrs	___
Controls	___
___	___
___	___
___	___
___	___

SUBCONTRACTORS ON PROJECT & THEIR ACTIVITY

DAILY ACTIVITY - VISITORS - INSPECTIONS

Supervisor's Signature: ______________

DAILY LOG

Day: ____________________

June 11, 20____ **Weather: 8:00 A.M. ___ Noon: ___ 4:00 P.M. ___**

WORK FORCE

Supt.	___
Foreman	___
Carpenters	___
Laborers	___
Masons	___
Oper, Engrs.	___
Iron Wrkers	___
Electricians	___
Plumbers	___
Steam Fittrs	___
Sheet Metal	___
Glazers	___
Roofers	___
Sprinkler	___
Painters	___
Tile Setters	___
Carpet Lyrs	___
Controls	___
___	___
___	___
___	___
___	___

SUBCONTRACTORS ON PROJECT & THEIR ACTIVITY

DAILY ACTIVITY - VISITORS - INSPECTIONS

Supervisor's Signature: ______________________________

DAILY LOG

Day: ____________________

June 12, 20____ **Weather: 8:00 A.M. ___ Noon: ___ 4:00 P.M. ___**

WORK FORCE

Supt.	___
Foreman	___
Carpenters	___
Laborers	___
Masons	___
Oper, Engrs.	___
Iron Wrkers	___
Electricians	___
Plumbers	___
Steam Fittrs	___
Sheet Metal	___
Glazers	___
Roofers	___
Sprinkler	___
Painters	___
Tile Setters	___
Carpet Lyrs	___
Controls	___
______	___
______	___
______	___
______	___

SUBCONTRACTORS ON PROJECT & THEIR ACTIVITY

DAILY ACTIVITY - VISITORS - INSPECTIONS

Supervisor's Signature: ______________________________

DAILY LOG

Day: ____________________

June 13, 20____ **Weather: 8:00 A.M. ___ Noon: ___ 4:00 P.M. ___**

WORK FORCE	
Supt.	____
Foreman	____
Carpenters	____
Laborers	____
Masons	____
Oper, Engrs.	____
Iron Wrkers	____
Electricians	____
Plumbers	____
Steam Fittrs	____
Sheet Metal	____
Glazers	____
Roofers	____
Sprinkler	____
Painters	____
Tile Setters	____
Carpet Lyrs	____
Controls	____
________	____
________	____
________	____
________	____

SUBCONTRACTORS ON PROJECT & THEIR ACTIVITY

DAILY ACTIVITY - VISITORS - INSPECTIONS

Supervisor's Signature: ____________________

DAILY LOG

Day: ____________________

June 14, 20____ **Weather: 8:00 A.M.** ___ **Noon:** ___ **4:00 P.M.** ___

WORK FORCE

Supt.	___
Foreman	___
Carpenters	___
Laborers	___
Masons	___
Oper, Engrs.	___
Iron Wrkers	___
Electricians	___
Plumbers	___
Steam Fittrs	___
Sheet Metal	___
Glazers	___
Roofers	___
Sprinkler	___
Painters	___
Tile Setters	___
Carpet Lyrs	___
Controls	___
___	___
___	___
___	___
___	___

SUBCONTRACTORS ON PROJECT & THEIR ACTIVITY

DAILY ACTIVITY - VISITORS - INSPECTIONS

Supervisor's Signature: ____________________

DAILY LOG

Day: ____________________

June 15, 20____ **Weather: 8:00 A.M. ___ Noon: ___ 4:00 P.M. ___**

WORK FORCE	
Supt.	___
Foreman	___
Carpenters	___
Laborers	___
Masons	___
Oper, Engrs.	___
Iron Wrkers	___
Electricians	___
Plumbers	___
Steam Fittrs	___
Sheet Metal	___
Glazers	___
Roofers	___
Sprinkler	___
Painters	___
Tile Setters	___
Carpet Lyrs	___
Controls	___
________	___
________	___
________	___
________	___

SUBCONTRACTORS ON PROJECT & THEIR ACTIVITY

DAILY ACTIVITY - VISITORS - INSPECTIONS

Supervisor's Signature: ____________________

DAILY LOG

Day: ________________

June 16, 20____ **Weather: 8:00 A.M. ___ Noon: ___ 4:00 P.M. ___**

WORK FORCE	
Supt.	___
Foreman	___
Carpenters	___
Laborers	___
Masons	___
Oper, Engrs.	___
Iron Wrkers	___
Electricians	___
Plumbers	___
Steam Fittrs	___
Sheet Metal	___
Glazers	___
Roofers	___
Sprinkler	___
Painters	___
Tile Setters	___
Carpet Lyrs	___
Controls	___
______	___
______	___
______	___
______	___

SUBCONTRACTORS ON PROJECT & THEIR ACTIVITY

DAILY ACTIVITY - VISITORS - INSPECTIONS

Supervisor's Signature: ______________________________

DAILY LOG

Day: ____________________

June 17, 20____

Weather: 8:00 A.M. ___ Noon: ___ 4:00 P.M. ___

WORK FORCE

Supt.	____
Foreman	____
Carpenters	____
Laborers	____
Masons	____
Oper, Engrs.	____
Iron Wrkers	____
Electricians	____
Plumbers	____
Steam Fittrs	____
Sheet Metal	____
Glazers	____
Roofers	____
Sprinkler	____
Painters	____
Tile Setters	____
Carpet Lyrs	____
Controls	____
________	____
________	____
________	____
________	____

SUBCONTRACTORS ON PROJECT & THEIR ACTIVITY

DAILY ACTIVITY - VISITORS - INSPECTIONS

Supervisor's Signature: ______________________________

DAILY LOG

Day: ________________

June 18, 20____ **Weather: 8:00 A.M. ___ Noon: ___ 4:00 P.M. ___**

WORK FORCE	
Supt.	____
Foreman	____
Carpenters	____
Laborers	____
Masons	____
Oper, Engrs.	____
Iron Wrkers	____
Electricians	____
Plumbers	____
Steam Fittrs	____
Sheet Metal	____
Glazers	____
Roofers	____
Sprinkler	____
Painters	____
Tile Setters	____
Carpet Lyrs	____
Controls	____
________	____
________	____
________	____
________	____

SUBCONTRACTORS ON PROJECT & THEIR ACTIVITY

DAILY ACTIVITY - VISITORS - INSPECTIONS

Supervisor's Signature: ____________________

DAILY LOG

Day: ____________________

June 19, 20____

Weather: 8:00 A.M. ___ Noon: ___ 4:00 P.M. ___

WORK FORCE

Supt.	____
Foreman	____
Carpenters	____
Laborers	____
Masons	____
Oper, Engrs.	____
Iron Wrkers	____
Electricians	____
Plumbers	____
Steam Fittrs	____
Sheet Metal	____
Glazers	____
Roofers	____
Sprinkler	____
Painters	____
Tile Setters	____
Carpet Lyrs	____
Controls	____
________	____
________	____
________	____
________	____

SUBCONTRACTORS ON PROJECT & THEIR ACTIVITY

DAILY ACTIVITY - VISITORS - INSPECTIONS

Supervisor's Signature: ______________________________

DAILY LOG

Day: ________________

June 20, 20____   **Weather: 8:00 A.M. ___ Noon: ___ 4:00 P.M. ___**

WORK FORCE

Supt.	___
Foreman	___
Carpenters	___
Laborers	___
Masons	___
Oper, Engrs.	___
Iron Wrkers	___
Electricians	___
Plumbers	___
Steam Fittrs	___
Sheet Metal	___
Glazers	___
Roofers	___
Sprinkler	___
Painters	___
Tile Setters	___
Carpet Lyrs	___
Controls	___
________	___
________	___
________	___
________	___

SUBCONTRACTORS ON PROJECT & THEIR ACTIVITY

DAILY ACTIVITY - VISITORS - INSPECTIONS

Supervisor's Signature: ________________________________

DAILY LOG

Day: ____________________

June 21, 20____ **Weather: 8:00 A.M. ___ Noon: ___ 4:00 P.M. ___**

WORK FORCE	
Supt.	____
Foreman	____
Carpenters	____
Laborers	____
Masons	____
Oper, Engrs.	____
Iron Wrkers	____
Electricians	____
Plumbers	____
Steam Fittrs	____
Sheet Metal	____
Glazers	____
Roofers	____
Sprinkler	____
Painters	____
Tile Setters	____
Carpet Lyrs	____
Controls	____
________	____
________	____
________	____
________	____

SUBCONTRACTORS ON PROJECT & THEIR ACTIVITY

DAILY ACTIVITY - VISITORS - INSPECTIONS

Supervisor's Signature: ______________________________

DAILY LOG

Day: ________________

June 22, 20____ **Weather: 8:00 A.M. ___ Noon: ___ 4:00 P.M. ___**

WORK FORCE

Supt.	___
Foreman	___
Carpenters	___
Laborers	___
Masons	___
Oper, Engrs.	___
Iron Wrkers	___
Electricians	___
Plumbers	___
Steam Fittrs	___
Sheet Metal	___
Glazers	___
Roofers	___
Sprinkler	___
Painters	___
Tile Setters	___
Carpet Lyrs	___
Controls	___
________	___
________	___
________	___
________	___

SUBCONTRACTORS ON PROJECT & THEIR ACTIVITY

DAILY ACTIVITY - VISITORS - INSPECTIONS

Supervisor's Signature: ______________________________

DAILY LOG

Day: ____________________

June 23, 20____

Weather: 8:00 A.M. ___ Noon: ___ 4:00 P.M. ___

WORK FORCE

Supt.	___
Foreman	___
Carpenters	___
Laborers	___
Masons	___
Oper, Engrs.	___
Iron Wrkers	___
Electricians	___
Plumbers	___
Steam Fittrs	___
Sheet Metal	___
Glazers	___
Roofers	___
Sprinkler	___
Painters	___
Tile Setters	___
Carpet Lyrs	___
Controls	___
______	___
______	___
______	___
______	___

SUBCONTRACTORS ON PROJECT & THEIR ACTIVITY

DAILY ACTIVITY - VISITORS - INSPECTIONS

Supervisor's Signature: ______________________________

DAILY LOG

Day: ____________________

June 24, 20____ **Weather: 8:00 A.M. ____ Noon: ____ 4:00 P.M. ____**

WORK FORCE

Supt.	____
Foreman	____
Carpenters	____
Laborers	____
Masons	____
Oper, Engrs.	____
Iron Wrkers	____
Electricians	____
Plumbers	____
Steam Fittrs	____
Sheet Metal	____
Glazers	____
Roofers	____
Sprinkler	____
Painters	____
Tile Setters	____
Carpet Lyrs	____
Controls	____
________	____
________	____
________	____
________	____

SUBCONTRACTORS ON PROJECT & THEIR ACTIVITY

DAILY ACTIVITY - VISITORS - INSPECTIONS

Supervisor's Signature: ______________________________

DAILY LOG

Day: ____________________

June 25, 20____ **Weather: 8:00 A.M. ____ Noon: ____ 4:00 P.M. ____**

WORK FORCE	
Supt.	____
Foreman	____
Carpenters	____
Laborers	____
Masons	____
Oper, Engrs.	____
Iron Wrkers	____
Electricians	____
Plumbers	____
Steam Fittrs	____
Sheet Metal	____
Glazers	____
Roofers	____
Sprinkler	____
Painters	____
Tile Setters	____
Carpet Lyrs	____
Controls	____
________	____
________	____
________	____
________	____

SUBCONTRACTORS ON PROJECT & THEIR ACTIVITY

DAILY ACTIVITY - VISITORS - INSPECTIONS

Supervisor's Signature: ______________________________

DAILY LOG

Day: ____________________

June 26, 20____

Weather: 8:00 A.M. ____ Noon: ____ 4:00 P.M. ____

WORK FORCE	
Supt.	____
Foreman	____
Carpenters	____
Laborers	____
Masons	____
Oper, Engrs.	____
Iron Wrkers	____
Electricians	____
Plumbers	____
Steam Fittrs	____
Sheet Metal	____
Glazers	____
Roofers	____
Sprinkler	____
Painters	____
Tile Setters	____
Carpet Lyrs	____
Controls	____
________	____
________	____
________	____
________	____

SUBCONTRACTORS ON PROJECT & THEIR ACTIVITY

DAILY ACTIVITY - VISITORS - INSPECTIONS

Supervisor's Signature: ______________________________

DAILY LOG

Day: ____________________

June 27, 20____ **Weather: 8:00 A.M. ___ Noon: ___ 4:00 P.M. ___**

WORK FORCE

Supt.	____
Foreman	____
Carpenters	____
Laborers	____
Masons	____
Oper, Engrs.	____
Iron Wrkers	____
Electricians	____
Plumbers	____
Steam Fittrs	____
Sheet Metal	____
Glazers	____
Roofers	____
Sprinkler	____
Painters	____
Tile Setters	____
Carpet Lyrs	____
Controls	____
________	____
________	____
________	____
________	____

SUBCONTRACTORS ON PROJECT & THEIR ACTIVITY

DAILY ACTIVITY - VISITORS - INSPECTIONS

Supervisor's Signature: ________________________________

DAILY LOG

Day: ________________

June 28, 20____ **Weather: 8:00 A.M. ___ Noon: ___ 4:00 P.M. ___**

WORK FORCE	
Supt.	___
Foreman	___
Carpenters	___
Laborers	___
Masons	___
Oper, Engrs.	___
Iron Wrkers	___
Electricians	___
Plumbers	___
Steam Fittrs	___
Sheet Metal	___
Glazers	___
Roofers	___
Sprinkler	___
Painters	___
Tile Setters	___
Carpet Lyrs	___
Controls	___
________	___
________	___
________	___
________	___

SUBCONTRACTORS ON PROJECT & THEIR ACTIVITY

DAILY ACTIVITY - VISITORS - INSPECTIONS

Supervisor's Signature: ______________________________

DAILY LOG

Day: ____________________

June 29, 20____

Weather: 8:00 A.M. ___ Noon: ___ 4:00 P.M. ___

WORK FORCE

Supt.	____
Foreman	____
Carpenters	____
Laborers	____
Masons	____
Oper, Engrs.	____
Iron Wrkers	____
Electricians	____
Plumbers	____
Steam Fittrs	____
Sheet Metal	____
Glazers	____
Roofers	____
Sprinkler	____
Painters	____
Tile Setters	____
Carpet Lyrs	____
Controls	____
____	____
____	____
____	____
____	____

SUBCONTRACTORS ON PROJECT & THEIR ACTIVITY

DAILY ACTIVITY - VISITORS - INSPECTIONS

Supervisor's Signature: ________________________________

DAILY LOG

Day: ____________________

June 30, 20____ **Weather: 8:00 A.M.** ___ **Noon:** ___ **4:00 P.M.** ___

WORK FORCE	
Supt.	___
Foreman	___
Carpenters	___
Laborers	___
Masons	___
Oper, Engrs.	___
Iron Wrkers	___
Electricians	___
Plumbers	___
Steam Fittrs	___
Sheet Metal	___
Glazers	___
Roofers	___
Sprinkler	___
Painters	___
Tile Setters	___
Carpet Lyrs	___
Controls	___
________	___
________	___
________	___
________	___

SUBCONTRACTORS ON PROJECT & THEIR ACTIVITY

DAILY ACTIVITY - VISITORS - INSPECTIONS

Supervisor's Signature: ____________________

DAILY LOG

Day: ____________________

July 1, 20____   **Weather: 8:00 A.M. ___ Noon: ___ 4:00 P.M. ___**

WORK FORCE		SUBCONTRACTORS ON PROJECT & THEIR ACTIVITY
Supt.	___	
Foreman	___	
Carpenters	___	
Laborers	___	
Masons	___	
Oper, Engrs.	___	
Iron Wrkers	___	
Electricians	___	
Plumbers	___	
Steam Fittrs	___	
Sheet Metal	___	
Glazers	___	
Roofers	___	
Sprinkler	___	
Painters	___	
Tile Setters	___	
Carpet Lyrs	___	
Controls	___	
___	___	DAILY ACTIVITY - VISITORS - INSPECTIONS
___	___	
___	___	
___	___	

Supervisor's Signature: ______________________________

DAILY LOG

Day: ____________________

July 2, 20____

Weather: 8:00 A.M. ____ Noon: ____ 4:00 P.M. ____

WORK FORCE

Supt.	____
Foreman	____
Carpenters	____
Laborers	____
Masons	____
Oper, Engrs.	____
Iron Wrkers	____
Electricians	____
Plumbers	____
Steam Fittrs	____
Sheet Metal	____
Glazers	____
Roofers	____
Sprinkler	____
Painters	____
Tile Setters	____
Carpet Lyrs	____
Controls	____
________	____
________	____
________	____
________	____

SUBCONTRACTORS ON PROJECT & THEIR ACTIVITY

DAILY ACTIVITY - VISITORS - INSPECTIONS

Supervisor's Signature: ________________________________

DAILY LOG

Day: ______________

July 3, 20___ **Weather: 8:00 A.M. ___ Noon: ___ 4:00 P.M. ___**

WORK FORCE

Supt.	___
Foreman	___
Carpenters	___
Laborers	___
Masons	___
Oper, Engrs.	___
Iron Wrkers	___
Electricians	___
Plumbers	___
Steam Fittrs	___
Sheet Metal	___
Glazers	___
Roofers	___
Sprinkler	___
Painters	___
Tile Setters	___
Carpet Lyrs	___
Controls	___
______	___
______	___
______	___
______	___

SUBCONTRACTORS ON PROJECT & THEIR ACTIVITY

DAILY ACTIVITY - VISITORS - INSPECTIONS

Supervisor's Signature: ______________________________

DAILY LOG

Day: ________________

July 4, 20___ **Weather: 8:00 A.M. ___ Noon: ___ 4:00 P.M. ___**

WORK FORCE

Supt.	___
Foreman	___
Carpenters	___
Laborers	___
Masons	___
Oper, Engrs.	___
Iron Wrkers	___
Electricians	___
Plumbers	___
Steam Fittrs	___
Sheet Metal	___
Glazers	___
Roofers	___
Sprinkler	___
Painters	___
Tile Setters	___
Carpet Lyrs	___
Controls	___
________	___
________	___
________	___
________	___

SUBCONTRACTORS ON PROJECT & THEIR ACTIVITY

DAILY ACTIVITY - VISITORS - INSPECTIONS

Supervisor's Signature: ________________________________

DAILY LOG

Day: ____________________

July 5, 20____ **Weather: 8:00 A.M. ____ Noon: ____ 4:00 P.M. ____**

WORK FORCE

Supt.	____
Foreman	____
Carpenters	____
Laborers	____
Masons	____
Oper, Engrs.	____
Iron Wrkers	____
Electricians	____
Plumbers	____
Steam Fittrs	____
Sheet Metal	____
Glazers	____
Roofers	____
Sprinkler	____
Painters	____
Tile Setters	____
Carpet Lyrs	____
Controls	____
________	____
________	____
________	____
________	____

SUBCONTRACTORS ON PROJECT & THEIR ACTIVITY

DAILY ACTIVITY - VISITORS - INSPECTIONS

Supervisor's Signature: ______________________________

DAILY LOG

Day: ____________________

July 6, 20____

Weather: 8:00 A.M. ____ Noon: ____ 4:00 P.M. ____

WORK FORCE

Supt.	____
Foreman	____
Carpenters	____
Laborers	____
Masons	____
Oper, Engrs.	____
Iron Wrkers	____
Electricians	____
Plumbers	____
Steam Fittrs	____
Sheet Metal	____
Glazers	____
Roofers	____
Sprinkler	____
Painters	____
Tile Setters	____
Carpet Lyrs	____
Controls	____
________	____
________	____
________	____
________	____

SUBCONTRACTORS ON PROJECT & THEIR ACTIVITY

DAILY ACTIVITY - VISITORS - INSPECTIONS

Supervisor's Signature: ______________________________

DAILY LOG

Day: ____________________

July 7, 20____ **Weather: 8:00 A.M. ___ Noon: ___ 4:00 P.M. ___**

WORK FORCE		SUBCONTRACTORS ON PROJECT & THEIR ACTIVITY
Supt.	____	
Foreman	____	
Carpenters	____	
Laborers	____	
Masons	____	
Oper, Engrs.	____	
Iron Wrkers	____	
Electricians	____	
Plumbers	____	
Steam Fittrs	____	
Sheet Metal	____	
Glazers	____	
Roofers	____	
Sprinkler	____	
Painters	____	
Tile Setters	____	
Carpet Lyrs	____	
Controls	____	
________	____	DAILY ACTIVITY - VISITORS - INSPECTIONS
________	____	
________	____	
________	____	

Supervisor's Signature: ______________________________

DAILY LOG

Day: ____________________

July 8, 20____

Weather: 8:00 A.M. ___ Noon: ___ 4:00 P.M. ___

WORK FORCE	
Supt.	____
Foreman	____
Carpenters	____
Laborers	____
Masons	____
Oper, Engrs.	____
Iron Wrkers	____
Electricians	____
Plumbers	____
Steam Fittrs	____
Sheet Metal	____
Glazers	____
Roofers	____
Sprinkler	____
Painters	____
Tile Setters	____
Carpet Lyrs	____
Controls	____
________	____
________	____
________	____
________	____

SUBCONTRACTORS ON PROJECT & THEIR ACTIVITY

DAILY ACTIVITY - VISITORS - INSPECTIONS

Supervisor's Signature: ______________________________

DAILY LOG

Day: ____________________

July 9, 20____

Weather: 8:00 A.M. ___ Noon: ___ 4:00 P.M. ___

WORK FORCE

Supt.	___
Foreman	___
Carpenters	___
Laborers	___
Masons	___
Oper, Engrs.	___
Iron Wrkers	___
Electricians	___
Plumbers	___
Steam Fittrs	___
Sheet Metal	___
Glazers	___
Roofers	___
Sprinkler	___
Painters	___
Tile Setters	___
Carpet Lyrs	___
Controls	___
______	___
______	___
______	___
______	___

SUBCONTRACTORS ON PROJECT & THEIR ACTIVITY

DAILY ACTIVITY - VISITORS - INSPECTIONS

Supervisor's Signature: ____________________

DAILY LOG

Day: ____________________

July 10, 20____

Weather: 8:00 A.M. ___ Noon: ___ 4:00 P.M. ___

WORK FORCE

Supt.	___
Foreman	___
Carpenters	___
Laborers	___
Masons	___
Oper, Engrs.	___
Iron Wrkers	___
Electricians	___
Plumbers	___
Steam Fittrs	___
Sheet Metal	___
Glazers	___
Roofers	___
Sprinkler	___
Painters	___
Tile Setters	___
Carpet Lyrs	___
Controls	___
______	___
______	___
______	___
______	___

SUBCONTRACTORS ON PROJECT & THEIR ACTIVITY

DAILY ACTIVITY - VISITORS - INSPECTIONS

Supervisor's Signature: ______________________________

DAILY LOG

Day: ____________________

July 11, 20____ **Weather: 8:00 A.M. ___ Noon: ___ 4:00 P.M. ___**

WORK FORCE

Supt.	___
Foreman	___
Carpenters	___
Laborers	___
Masons	___
Oper, Engrs.	___
Iron Wrkers	___
Electricians	___
Plumbers	___
Steam Fittrs	___
Sheet Metal	___
Glazers	___
Roofers	___
Sprinkler	___
Painters	___
Tile Setters	___
Carpet Lyrs	___
Controls	___
______	___
______	___
______	___
______	___

SUBCONTRACTORS ON PROJECT & THEIR ACTIVITY

DAILY ACTIVITY - VISITORS - INSPECTIONS

Supervisor's Signature: ____________________

DAILY LOG

Day: ____________________

July 12, 20____ **Weather: 8:00 A.M. ___ Noon: ___ 4:00 P.M. ___**

WORK FORCE

Supt.	____
Foreman	____
Carpenters	____
Laborers	____
Masons	____
Oper, Engrs.	____
Iron Wrkers	____
Electricians	____
Plumbers	____
Steam Fittrs	____
Sheet Metal	____
Glazers	____
Roofers	____
Sprinkler	____
Painters	____
Tile Setters	____
Carpet Lyrs	____
Controls	____
________	____
________	____
________	____
________	____

SUBCONTRACTORS ON PROJECT & THEIR ACTIVITY

DAILY ACTIVITY - VISITORS - INSPECTIONS

Supervisor's Signature: ______________________________

DAILY LOG

Day: ____________________

July 13, 20____ **Weather: 8:00 A.M. ___ Noon: ___ 4:00 P.M. ___**

WORK FORCE	
Supt.	____
Foreman	____
Carpenters	____
Laborers	____
Masons	____
Oper, Engrs.	____
Iron Wrkers	____
Electricians	____
Plumbers	____
Steam Fittrs	____
Sheet Metal	____
Glazers	____
Roofers	____
Sprinkler	____
Painters	____
Tile Setters	____
Carpet Lyrs	____
Controls	____
________	____
________	____
________	____
________	____

SUBCONTRACTORS ON PROJECT & THEIR ACTIVITY

DAILY ACTIVITY - VISITORS - INSPECTIONS

Supervisor's Signature: ______________________________

DAILY LOG

Day: ____________________

July 14, 20____ **Weather: 8:00 A.M. ___ Noon: ___ 4:00 P.M. ___**

WORK FORCE **SUBCONTRACTORS ON PROJECT & THEIR ACTIVITY**

WORK FORCE	
Supt.	____
Foreman	____
Carpenters	____
Laborers	____
Masons	____
Oper, Engrs.	____
Iron Wrkers	____
Electricians	____
Plumbers	____
Steam Fittrs	____
Sheet Metal	____
Glazers	____
Roofers	____
Sprinkler	____
Painters	____
Tile Setters	____
Carpet Lyrs	____
Controls	____
________	____
________	____
________	____
________	____

DAILY ACTIVITY - VISITORS - INSPECTIONS

Supervisor's Signature: ______________________________

DAILY LOG

Day: ____________________

July 15, 20____   **Weather: 8:00 A.M. ___ Noon: ___ 4:00 P.M. ___**

WORK FORCE

Supt.	___
Foreman	___
Carpenters	___
Laborers	___
Masons	___
Oper, Engrs.	___
Iron Wrkers	___
Electricians	___
Plumbers	___
Steam Fittrs	___
Sheet Metal	___
Glazers	___
Roofers	___
Sprinkler	___
Painters	___
Tile Setters	___
Carpet Lyrs	___
Controls	___
______	___
______	___
______	___
______	___

SUBCONTRACTORS ON PROJECT & THEIR ACTIVITY

DAILY ACTIVITY - VISITORS - INSPECTIONS

Supervisor's Signature: ______________________________

DAILY LOG

Day: ____________________

July 16, 20____ **Weather: 8:00 A.M. ___ Noon: ___ 4:00 P.M. ___**

WORK FORCE		SUBCONTRACTORS ON PROJECT & THEIR ACTIVITY
Supt.	____	
Foreman	____	
Carpenters	____	
Laborers	____	
Masons	____	
Oper, Engrs.	____	
Iron Wrkers	____	
Electricians	____	
Plumbers	____	
Steam Fittrs	____	
Sheet Metal	____	
Glazers	____	
Roofers	____	
Sprinkler	____	
Painters	____	
Tile Setters	____	
Carpet Lyrs	____	
Controls	____	
________	____	DAILY ACTIVITY - VISITORS - INSPECTIONS
________	____	
________	____	
________	____	

Supervisor's Signature: ______________________________

DAILY LOG

Day: ____________________

July 17, 20____ **Weather: 8:00 A.M. ___ Noon: ___ 4:00 P.M. ___**

WORK FORCE

Supt.	____
Foreman	____
Carpenters	____
Laborers	____
Masons	____
Oper, Engrs.	____
Iron Wrkers	____
Electricians	____
Plumbers	____
Steam Fittrs	____
Sheet Metal	____
Glazers	____
Roofers	____
Sprinkler	____
Painters	____
Tile Setters	____
Carpet Lyrs	____
Controls	____
________	____
________	____
________	____
________	____

SUBCONTRACTORS ON PROJECT & THEIR ACTIVITY

DAILY ACTIVITY - VISITORS - INSPECTIONS

Supervisor's Signature: ______________________________

DAILY LOG

Day: ____________________

July 18, 20____

Weather: 8:00 A.M. ___ Noon: ___ 4:00 P.M. ___

WORK FORCE

Supt.	___
Foreman	___
Carpenters	___
Laborers	___
Masons	___
Oper, Engrs.	___
Iron Wrkers	___
Electricians	___
Plumbers	___
Steam Fittrs	___
Sheet Metal	___
Glazers	___
Roofers	___
Sprinkler	___
Painters	___
Tile Setters	___
Carpet Lyrs	___
Controls	___
________	___
________	___
________	___
________	___

SUBCONTRACTORS ON PROJECT & THEIR ACTIVITY

DAILY ACTIVITY - VISITORS - INSPECTIONS

Supervisor's Signature: ______________________________

DAILY LOG

Day: ____________________

July 19, 20____   **Weather: 8:00 A.M. ____ Noon: ____ 4:00 P.M. ____**

WORK FORCE

Supt.	____
Foreman	____
Carpenters	____
Laborers	____
Masons	____
Oper, Engrs.	____
Iron Wrkers	____
Electricians	____
Plumbers	____
Steam Fittrs	____
Sheet Metal	____
Glazers	____
Roofers	____
Sprinkler	____
Painters	____
Tile Setters	____
Carpet Lyrs	____
Controls	____
________	____
________	____
________	____
________	____

SUBCONTRACTORS ON PROJECT & THEIR ACTIVITY

DAILY ACTIVITY - VISITORS - INSPECTIONS

Supervisor's Signature: ______________________________

DAILY LOG

Day: ____________________

July 20, 20____ **Weather: 8:00 A.M. ___ Noon: ___ 4:00 P.M. ___**

WORK FORCE

Supt.	___
Foreman	___
Carpenters	___
Laborers	___
Masons	___
Oper, Engrs.	___
Iron Wrkers	___
Electricians	___
Plumbers	___
Steam Fittrs	___
Sheet Metal	___
Glazers	___
Roofers	___
Sprinkler	___
Painters	___
Tile Setters	___
Carpet Lyrs	___
Controls	___
______	___
______	___
______	___
______	___

SUBCONTRACTORS ON PROJECT & THEIR ACTIVITY

DAILY ACTIVITY - VISITORS - INSPECTIONS

Supervisor's Signature: ______________________________

DAILY LOG

Day: ____________________

July 21, 20____ **Weather: 8:00 A.M. ___ Noon: ___ 4:00 P.M. ___**

WORK FORCE

Supt.	____
Foreman	____
Carpenters	____
Laborers	____
Masons	____
Oper, Engrs.	____
Iron Wrkers	____
Electricians	____
Plumbers	____
Steam Fittrs	____
Sheet Metal	____
Glazers	____
Roofers	____
Sprinkler	____
Painters	____
Tile Setters	____
Carpet Lyrs	____
Controls	____
________	____
________	____
________	____
________	____

SUBCONTRACTORS ON PROJECT & THEIR ACTIVITY

DAILY ACTIVITY - VISITORS - INSPECTIONS

Supervisor's Signature: ________________________________

DAILY LOG

Day: ____________________

July 22, 20____   **Weather: 8:00 A.M. ____ Noon: ____ 4:00 P.M. ____**

WORK FORCE

Supt.	____
Foreman	____
Carpenters	____
Laborers	____
Masons	____
Oper, Engrs.	____
Iron Wrkers	____
Electricians	____
Plumbers	____
Steam Fittrs	____
Sheet Metal	____
Glazers	____
Roofers	____
Sprinkler	____
Painters	____
Tile Setters	____
Carpet Lyrs	____
Controls	____
________	____
________	____
________	____
________	____

SUBCONTRACTORS ON PROJECT & THEIR ACTIVITY

DAILY ACTIVITY - VISITORS - INSPECTIONS

Supervisor's Signature: ______________________________

DAILY LOG

Day: ____________________

July 23, 20____ **Weather: 8:00 A.M. ____ Noon: ____ 4:00 P.M. ____**

WORK FORCE		**SUBCONTRACTORS ON PROJECT & THEIR ACTIVITY**
Supt.	____	
Foreman	____	
Carpenters	____	
Laborers	____	
Masons	____	
Oper, Engrs.	____	
Iron Wrkers	____	
Electricians	____	
Plumbers	____	
Steam Fittrs	____	
Sheet Metal	____	
Glazers	____	
Roofers	____	
Sprinkler	____	
Painters	____	
Tile Setters	____	
Carpet Lyrs	____	
Controls	____	
________	____	DAILY ACTIVITY - VISITORS - INSPECTIONS
________	____	
________	____	
________	____	

Supervisor's Signature: ______________________________

DAILY LOG

Day: ____________________

July 24, 20____ **Weather: 8:00 A.M. ____ Noon: ____ 4:00 P.M. ____**

WORK FORCE | **SUBCONTRACTORS ON PROJECT & THEIR ACTIVITY**

WORK FORCE	
Supt.	
Foreman	
Carpenters	
Laborers	
Masons	
Oper, Engrs.	
Iron Wrkers	
Electricians	
Plumbers	
Steam Fittrs	
Sheet Metal	
Glazers	
Roofers	
Sprinkler	
Painters	
Tile Setters	
Carpet Lyrs	
Controls	

DAILY ACTIVITY - VISITORS - INSPECTIONS

Supervisor's Signature: ____________________

DAILY LOG

Day: ____________________

July 25, 20____ **Weather: 8:00 A.M. ___ Noon: ___ 4:00 P.M. ___**

WORK FORCE	
Supt.	____
Foreman	____
Carpenters	____
Laborers	____
Masons	____
Oper, Engrs.	____
Iron Wrkers	____
Electricians	____
Plumbers	____
Steam Fittrs	____
Sheet Metal	____
Glazers	____
Roofers	____
Sprinkler	____
Painters	____
Tile Setters	____
Carpet Lyrs	____
Controls	____
________	____
________	____
________	____
________	____

SUBCONTRACTORS ON PROJECT & THEIR ACTIVITY

DAILY ACTIVITY - VISITORS - INSPECTIONS

Supervisor's Signature: ______________________________

DAILY LOG

Day: ____________________

July 26, 20____ **Weather: 8:00 A.M. ___ Noon: ___ 4:00 P.M. ___**

WORK FORCE

Supt.	____
Foreman	____
Carpenters	____
Laborers	____
Masons	____
Oper, Engrs.	____
Iron Wrkers	____
Electricians	____
Plumbers	____
Steam Fittrs	____
Sheet Metal	____
Glazers	____
Roofers	____
Sprinkler	____
Painters	____
Tile Setters	____
Carpet Lyrs	____
Controls	____
________	____
________	____
________	____
________	____

SUBCONTRACTORS ON PROJECT & THEIR ACTIVITY

DAILY ACTIVITY - VISITORS - INSPECTIONS

Supervisor's Signature: ______________________________

DAILY LOG

Day: ________________

July 27, 20___ **Weather: 8:00 A.M. ___ Noon: ___ 4:00 P.M. ___**

WORK FORCE

Supt.	___
Foreman	___
Carpenters	___
Laborers	___
Masons	___
Oper, Engrs.	___
Iron Wrkers	___
Electricians	___
Plumbers	___
Steam Fittrs	___
Sheet Metal	___
Glazers	___
Roofers	___
Sprinkler	___
Painters	___
Tile Setters	___
Carpet Lyrs	___
Controls	___
______	___
______	___
______	___
______	___

SUBCONTRACTORS ON PROJECT & THEIR ACTIVITY

DAILY ACTIVITY - VISITORS - INSPECTIONS

Supervisor's Signature: ________________

DAILY LOG

Day: ____________________

July 28, 20____ **Weather: 8:00 A.M. ___ Noon: ___ 4:00 P.M. ___**

WORK FORCE

Supt.	___
Foreman	___
Carpenters	___
Laborers	___
Masons	___
Oper, Engrs.	___
Iron Wrkers	___
Electricians	___
Plumbers	___
Steam Fittrs	___
Sheet Metal	___
Glazers	___
Roofers	___
Sprinkler	___
Painters	___
Tile Setters	___
Carpet Lyrs	___
Controls	___
______	___
______	___
______	___
______	___

SUBCONTRACTORS ON PROJECT & THEIR ACTIVITY

DAILY ACTIVITY - VISITORS - INSPECTIONS

Supervisor's Signature: ____________________

DAILY LOG

Day: ____________________

July 29, 20____ **Weather: 8:00 A.M. ___ Noon: ___ 4:00 P.M. ___**

WORK FORCE

Supt.	____
Foreman	____
Carpenters	____
Laborers	____
Masons	____
Oper, Engrs.	____
Iron Wrkers	____
Electricians	____
Plumbers	____
Steam Fittrs	____
Sheet Metal	____
Glazers	____
Roofers	____
Sprinkler	____
Painters	____
Tile Setters	____
Carpet Lyrs	____
Controls	____
________	____
________	____
________	____
________	____

SUBCONTRACTORS ON PROJECT & THEIR ACTIVITY

DAILY ACTIVITY - VISITORS - INSPECTIONS

Supervisor's Signature: ____________________

DAILY LOG

Day: ____________________

July 30, 20____ **Weather: 8:00 A.M. ___ Noon: ___ 4:00 P.M. ___**

WORK FORCE

Supt.	___
Foreman	___
Carpenters	___
Laborers	___
Masons	___
Oper, Engrs.	___
Iron Wrkers	___
Electricians	___
Plumbers	___
Steam Fittrs	___
Sheet Metal	___
Glazers	___
Roofers	___
Sprinkler	___
Painters	___
Tile Setters	___
Carpet Lyrs	___
Controls	___
______	___
______	___
______	___
______	___

SUBCONTRACTORS ON PROJECT & THEIR ACTIVITY

DAILY ACTIVITY - VISITORS - INSPECTIONS

Supervisor's Signature: ______________________________

DAILY LOG

Day: ____________________

July 31, 20____ **Weather: 8:00 A.M. ___ Noon: ___ 4:00 P.M. ___**

WORK FORCE	
Supt.	___
Foreman	___
Carpenters	___
Laborers	___
Masons	___
Oper, Engrs.	___
Iron Wrkers	___
Electricians	___
Plumbers	___
Steam Fittrs	___
Sheet Metal	___
Glazers	___
Roofers	___
Sprinkler	___
Painters	___
Tile Setters	___
Carpet Lyrs	___
Controls	___
___	___
___	___
___	___
___	___

SUBCONTRACTORS ON PROJECT & THEIR ACTIVITY

DAILY ACTIVITY - VISITORS - INSPECTIONS

Supervisor's Signature: ____________________

DAILY LOG

Day: ____________________

August 1, 20____ **Weather: 8:00 A.M. ___ Noon: ___ 4:00 P.M. ___**

WORK FORCE

Supt.	____
Foreman	____
Carpenters	____
Laborers	____
Masons	____
Oper, Engrs.	____
Iron Wrkers	____
Electricians	____
Plumbers	____
Steam Fittrs	____
Sheet Metal	____
Glazers	____
Roofers	____
Sprinkler	____
Painters	____
Tile Setters	____
Carpet Lyrs	____
Controls	____
________	____
________	____
________	____
________	____

SUBCONTRACTORS ON PROJECT & THEIR ACTIVITY

DAILY ACTIVITY - VISITORS - INSPECTIONS

Supervisor's Signature: ______________________________

DAILY LOG

Day: ______________

August 2, 20___ **Weather: 8:00 A.M.** ___ **Noon:** ___ **4:00 P.M.** ___

WORK FORCE	
Supt.	___
Foreman	___
Carpenters	___
Laborers	___
Masons	___
Oper, Engrs.	___
Iron Wrkers	___
Electricians	___
Plumbers	___
Steam Fittrs	___
Sheet Metal	___
Glazers	___
Roofers	___
Sprinkler	___
Painters	___
Tile Setters	___
Carpet Lyrs	___
Controls	___
___	___
___	___
___	___
___	___

SUBCONTRACTORS ON PROJECT & THEIR ACTIVITY

DAILY ACTIVITY - VISITORS - INSPECTIONS

Supervisor's Signature: ______________

DAILY LOG

Day: ________________

August 3, 20____ **Weather: 8:00 A.M. ___ Noon: ___ 4:00 P.M. ___**

WORK FORCE	
Supt.	___
Foreman	___
Carpenters	___
Laborers	___
Masons	___
Oper, Engrs.	___
Iron Wrkers	___
Electricians	___
Plumbers	___
Steam Fittrs	___
Sheet Metal	___
Glazers	___
Roofers	___
Sprinkler	___
Painters	___
Tile Setters	___
Carpet Lyrs	___
Controls	___
________	___
________	___
________	___
________	___

SUBCONTRACTORS ON PROJECT & THEIR ACTIVITY

DAILY ACTIVITY - VISITORS - INSPECTIONS

Supervisor's Signature: ____________________

DAILY LOG

Day: ____________________

August 4, 20____ **Weather: 8:00 A.M. ___ Noon: ___ 4:00 P.M. ___**

WORK FORCE	
Supt.	____
Foreman	____
Carpenters	____
Laborers	____
Masons	____
Oper, Engrs.	____
Iron Wrkers	____
Electricians	____
Plumbers	____
Steam Fittrs	____
Sheet Metal	____
Glazers	____
Roofers	____
Sprinkler	____
Painters	____
Tile Setters	____
Carpet Lyrs	____
Controls	____
________	____
________	____
________	____
________	____

SUBCONTRACTORS ON PROJECT & THEIR ACTIVITY

DAILY ACTIVITY - VISITORS - INSPECTIONS

Supervisor's Signature: ____

DAILY LOG

Day: ____________________

August 5, 20___ **Weather: 8:00 A.M. ___ Noon: ___ 4:00 P.M. ___**

WORK FORCE	
Supt.	___
Foreman	___
Carpenters	___
Laborers	___
Masons	___
Oper, Engrs.	___
Iron Wrkers	___
Electricians	___
Plumbers	___
Steam Fittrs	___
Sheet Metal	___
Glazers	___
Roofers	___
Sprinkler	___
Painters	___
Tile Setters	___
Carpet Lyrs	___
Controls	___
______	___
______	___
______	___
______	___

SUBCONTRACTORS ON PROJECT & THEIR ACTIVITY

DAILY ACTIVITY - VISITORS - INSPECTIONS

Supervisor's Signature: ____________________

DAILY LOG

Day: ____________________

August 6, 20____ **Weather: 8:00 A.M. ___ Noon: ___ 4:00 P.M. ___**

WORK FORCE	
Supt.	____
Foreman	____
Carpenters	____
Laborers	____
Masons	____
Oper, Engrs.	____
Iron Wrkers	____
Electricians	____
Plumbers	____
Steam Fittrs	____
Sheet Metal	____
Glazers	____
Roofers	____
Sprinkler	____
Painters	____
Tile Setters	____
Carpet Lyrs	____
Controls	____
________	____
________	____
________	____
________	____

SUBCONTRACTORS ON PROJECT & THEIR ACTIVITY

DAILY ACTIVITY - VISITORS - INSPECTIONS

Supervisor's Signature: ______________________________

DAILY LOG

Day: ____________________

August 7, 20____ **Weather: 8:00 A.M. ___ Noon: ___ 4:00 P.M. ___**

WORK FORCE

Supt.	___
Foreman	___
Carpenters	___
Laborers	___
Masons	___
Oper, Engrs.	___
Iron Wrkers	___
Electricians	___
Plumbers	___
Steam Fittrs	___
Sheet Metal	___
Glazers	___
Roofers	___
Sprinkler	___
Painters	___
Tile Setters	___
Carpet Lyrs	___
Controls	___
________	___
________	___
________	___
________	___

SUBCONTRACTORS ON PROJECT & THEIR ACTIVITY

DAILY ACTIVITY - VISITORS - INSPECTIONS

Supervisor's Signature: ______________________________

DAILY LOG

Day: ____________________

August 8, 20____ **Weather: 8:00 A.M. ____ Noon: ____ 4:00 P.M. ____**

WORK FORCE

Supt.	____
Foreman	____
Carpenters	____
Laborers	____
Masons	____
Oper, Engrs.	____
Iron Wrkers	____
Electricians	____
Plumbers	____
Steam Fittrs	____
Sheet Metal	____
Glazers	____
Roofers	____
Sprinkler	____
Painters	____
Tile Setters	____
Carpet Lyrs	____
Controls	____
________	____
________	____
________	____
________	____

SUBCONTRACTORS ON PROJECT & THEIR ACTIVITY

DAILY ACTIVITY - VISITORS - INSPECTIONS

Supervisor's Signature: ______________________________

DAILY LOG

Day: ____________________

August 9, 20____   **Weather: 8:00 A.M. ___ Noon: ___ 4:00 P.M. ___**

WORK FORCE	
Supt.	___
Foreman	___
Carpenters	___
Laborers	___
Masons	___
Oper, Engrs.	___
Iron Wrkers	___
Electricians	___
Plumbers	___
Steam Fittrs	___
Sheet Metal	___
Glazers	___
Roofers	___
Sprinkler	___
Painters	___
Tile Setters	___
Carpet Lyrs	___
Controls	___
______	___
______	___
______	___
______	___

SUBCONTRACTORS ON PROJECT & THEIR ACTIVITY

DAILY ACTIVITY - VISITORS - INSPECTIONS

Supervisor's Signature: ____________________

DAILY LOG

Day: ____________________

August 10, 20____ **Weather: 8:00 A.M. ___ Noon: ___ 4:00 P.M. ___**

WORK FORCE

Supt.	___
Foreman	___
Carpenters	___
Laborers	___
Masons	___
Oper, Engrs.	___
Iron Wrkers	___
Electricians	___
Plumbers	___
Steam Fittrs	___
Sheet Metal	___
Glazers	___
Roofers	___
Sprinkler	___
Painters	___
Tile Setters	___
Carpet Lyrs	___
Controls	___
___	___
___	___
___	___
___	___

SUBCONTRACTORS ON PROJECT & THEIR ACTIVITY

DAILY ACTIVITY - VISITORS - INSPECTIONS

Supervisor's Signature: ____________________

DAILY LOG

Day: ____________________

August 11, 20____ **Weather: 8:00 A.M. ___ Noon: ___ 4:00 P.M. ___**

WORK FORCE	
Supt.	___
Foreman	___
Carpenters	___
Laborers	___
Masons	___
Oper, Engrs.	___
Iron Wrkers	___
Electricians	___
Plumbers	___
Steam Fittrs	___
Sheet Metal	___
Glazers	___
Roofers	___
Sprinkler	___
Painters	___
Tile Setters	___
Carpet Lyrs	___
Controls	___
______	___
______	___
______	___
______	___

SUBCONTRACTORS ON PROJECT & THEIR ACTIVITY

DAILY ACTIVITY - VISITORS - INSPECTIONS

Supervisor's Signature: ______________________________

DAILY LOG

Day: ____________________

August 12, 20____ **Weather: 8:00 A.M. ___ Noon: ___ 4:00 P.M. ___**

WORK FORCE

Supt.	____
Foreman	____
Carpenters	____
Laborers	____
Masons	____
Oper, Engrs.	____
Iron Wrkers	____
Electricians	____
Plumbers	____
Steam Fittrs	____
Sheet Metal	____
Glazers	____
Roofers	____
Sprinkler	____
Painters	____
Tile Setters	____
Carpet Lyrs	____
Controls	____
________	____
________	____
________	____
________	____

SUBCONTRACTORS ON PROJECT & THEIR ACTIVITY

DAILY ACTIVITY - VISITORS - INSPECTIONS

Supervisor's Signature: ______________________________

DAILY LOG

Day: ____________________

August 13, 20____ **Weather: 8:00 A.M. ___ Noon: ___ 4:00 P.M. ___**

WORK FORCE

Supt.	___
Foreman	___
Carpenters	___
Laborers	___
Masons	___
Oper, Engrs.	___
Iron Wrkers	___
Electricians	___
Plumbers	___
Steam Fittrs	___
Sheet Metal	___
Glazers	___
Roofers	___
Sprinkler	___
Painters	___
Tile Setters	___
Carpet Lyrs	___
Controls	___
________	___
________	___
________	___
________	___

SUBCONTRACTORS ON PROJECT & THEIR ACTIVITY

DAILY ACTIVITY - VISITORS - INSPECTIONS

Supervisor's Signature: ______________________________

DAILY LOG

Day: ____________________

August 14, 20____ **Weather: 8:00 A.M. ___ Noon: ___ 4:00 P.M. ___**

WORK FORCE

Supt.	____
Foreman	____
Carpenters	____
Laborers	____
Masons	____
Oper, Engrs.	____
Iron Wrkers	____
Electricians	____
Plumbers	____
Steam Fittrs	____
Sheet Metal	____
Glazers	____
Roofers	____
Sprinkler	____
Painters	____
Tile Setters	____
Carpet Lyrs	____
Controls	____
________	____
________	____
________	____
________	____

SUBCONTRACTORS ON PROJECT & THEIR ACTIVITY

DAILY ACTIVITY - VISITORS - INSPECTIONS

Supervisor's Signature: ________________________________

DAILY LOG

Day: ____________________

August 15, 20____ **Weather: 8:00 A.M. ___ Noon: ___ 4:00 P.M. ___**

WORK FORCE

Supt.	____
Foreman	____
Carpenters	____
Laborers	____
Masons	____
Oper, Engrs.	____
Iron Wrkers	____
Electricians	____
Plumbers	____
Steam Fittrs	____
Sheet Metal	____
Glazers	____
Roofers	____
Sprinkler	____
Painters	____
Tile Setters	____
Carpet Lyrs	____
Controls	____
________	____
________	____
________	____
________	____

SUBCONTRACTORS ON PROJECT & THEIR ACTIVITY

DAILY ACTIVITY - VISITORS - INSPECTIONS

Supervisor's Signature: ______________________________

DAILY LOG

Day: ____________________

August 16, 20____ **Weather: 8:00 A.M. ___ Noon: ___ 4:00 P.M. ___**

WORK FORCE

Supt.	____
Foreman	____
Carpenters	____
Laborers	____
Masons	____
Oper, Engrs.	____
Iron Wrkers	____
Electricians	____
Plumbers	____
Steam Fittrs	____
Sheet Metal	____
Glazers	____
Roofers	____
Sprinkler	____
Painters	____
Tile Setters	____
Carpet Lyrs	____
Controls	____
________	____
________	____
________	____
________	____

SUBCONTRACTORS ON PROJECT & THEIR ACTIVITY

DAILY ACTIVITY - VISITORS - INSPECTIONS

Supervisor's Signature: ________________________________

DAILY LOG

Day: ____________________

August 17, 20____ **Weather: 8:00 A.M. ___ Noon: ___ 4:00 P.M. ___**

WORK FORCE

Supt.	____
Foreman	____
Carpenters	____
Laborers	____
Masons	____
Oper, Engrs.	____
Iron Wrkers	____
Electricians	____
Plumbers	____
Steam Fittrs	____
Sheet Metal	____
Glazers	____
Roofers	____
Sprinkler	____
Painters	____
Tile Setters	____
Carpet Lyrs	____
Controls	____
________	____
________	____
________	____
________	____

SUBCONTRACTORS ON PROJECT & THEIR ACTIVITY

DAILY ACTIVITY - VISITORS - INSPECTIONS

Supervisor's Signature: ______________________________

DAILY LOG

Day: ____________________

August 18, 20____ **Weather: 8:00 A.M. ___ Noon: ___ 4:00 P.M. ___**

WORK FORCE	
Supt.	____
Foreman	____
Carpenters	____
Laborers	____
Masons	____
Oper, Engrs.	____
Iron Wrkers	____
Electricians	____
Plumbers	____
Steam Fittrs	____
Sheet Metal	____
Glazers	____
Roofers	____
Sprinkler	____
Painters	____
Tile Setters	____
Carpet Lyrs	____
Controls	____
____	____
____	____
____	____
____	____

SUBCONTRACTORS ON PROJECT & THEIR ACTIVITY

DAILY ACTIVITY - VISITORS - INSPECTIONS

Supervisor's Signature: ______________________________

DAILY LOG

Day: ____________________

August 19, 20____ **Weather: 8:00 A.M. ___ Noon: ___ 4:00 P.M. ___**

WORK FORCE	
Supt.	___
Foreman	___
Carpenters	___
Laborers	___
Masons	___
Oper, Engrs.	___
Iron Wrkers	___
Electricians	___
Plumbers	___
Steam Fittrs	___
Sheet Metal	___
Glazers	___
Roofers	___
Sprinkler	___
Painters	___
Tile Setters	___
Carpet Lyrs	___
Controls	___
______	___
______	___
______	___
______	___

SUBCONTRACTORS ON PROJECT & THEIR ACTIVITY

DAILY ACTIVITY - VISITORS - INSPECTIONS

Supervisor's Signature: ____________________

DAILY LOG

Day: ____________________

August 20, 20____ **Weather: 8:00 A.M. ___ Noon: ___ 4:00 P.M. ___**

WORK FORCE

Supt.	___
Foreman	___
Carpenters	___
Laborers	___
Masons	___
Oper, Engrs.	___
Iron Wrkers	___
Electricians	___
Plumbers	___
Steam Fittrs	___
Sheet Metal	___
Glazers	___
Roofers	___
Sprinkler	___
Painters	___
Tile Setters	___
Carpet Lyrs	___
Controls	___
______	___
______	___
______	___
______	___

SUBCONTRACTORS ON PROJECT & THEIR ACTIVITY

DAILY ACTIVITY - VISITORS - INSPECTIONS

Supervisor's Signature: ______________________________

DAILY LOG

Day: ____________________

August 21, 20____ **Weather: 8:00 A.M. ____ Noon: ____ 4:00 P.M. ____**

WORK FORCE

Supt.	____
Foreman	____
Carpenters	____
Laborers	____
Masons	____
Oper, Engrs.	____
Iron Wrkers	____
Electricians	____
Plumbers	____
Steam Fittrs	____
Sheet Metal	____
Glazers	____
Roofers	____
Sprinkler	____
Painters	____
Tile Setters	____
Carpet Lyrs	____
Controls	____
________	____
________	____
________	____
________	____

SUBCONTRACTORS ON PROJECT & THEIR ACTIVITY

DAILY ACTIVITY - VISITORS - INSPECTIONS

Supervisor's Signature: ________________________

DAILY LOG

Day: ____________________

August 22, 20____ **Weather: 8:00 A.M. ___ Noon: ___ 4:00 P.M. ___**

WORK FORCE

Supt.	____
Foreman	____
Carpenters	____
Laborers	____
Masons	____
Oper, Engrs.	____
Iron Wrkers	____
Electricians	____
Plumbers	____
Steam Fittrs	____
Sheet Metal	____
Glazers	____
Roofers	____
Sprinkler	____
Painters	____
Tile Setters	____
Carpet Lyrs	____
Controls	____
________	____
________	____
________	____
________	____

SUBCONTRACTORS ON PROJECT & THEIR ACTIVITY

DAILY ACTIVITY - VISITORS - INSPECTIONS

Supervisor's Signature: ______________________________

DAILY LOG

Day: ____________________

August 23, 20____ **Weather: 8:00 A.M. ____ Noon: ____ 4:00 P.M. ____**

WORK FORCE	
Supt.	____
Foreman	____
Carpenters	____
Laborers	____
Masons	____
Oper, Engrs.	____
Iron Wrkers	____
Electricians	____
Plumbers	____
Steam Fittrs	____
Sheet Metal	____
Glazers	____
Roofers	____
Sprinkler	____
Painters	____
Tile Setters	____
Carpet Lyrs	____
Controls	____
________	____
________	____
________	____
________	____

SUBCONTRACTORS ON PROJECT & THEIR ACTIVITY

DAILY ACTIVITY - VISITORS - INSPECTIONS

Supervisor's Signature: ______________________________

DAILY LOG

Day: ____________________

August 24, 20____ **Weather: 8:00 A.M. ____ Noon: ____ 4:00 P.M. ____**

WORK FORCE

Supt.	____
Foreman	____
Carpenters	____
Laborers	____
Masons	____
Oper, Engrs.	____
Iron Wrkers	____
Electricians	____
Plumbers	____
Steam Fittrs	____
Sheet Metal	____
Glazers	____
Roofers	____
Sprinkler	____
Painters	____
Tile Setters	____
Carpet Lyrs	____
Controls	____
________	____
________	____
________	____
________	____

SUBCONTRACTORS ON PROJECT & THEIR ACTIVITY

DAILY ACTIVITY - VISITORS - INSPECTIONS

Supervisor's Signature: ____________________

DAILY LOG

Day: ____________________

August 25, 20____ **Weather: 8:00 A.M. ___ Noon: ___ 4:00 P.M. ___**

WORK FORCE	
Supt.	___
Foreman	___
Carpenters	___
Laborers	___
Masons	___
Oper, Engrs.	___
Iron Wrkers	___
Electricians	___
Plumbers	___
Steam Fittrs	___
Sheet Metal	___
Glazers	___
Roofers	___
Sprinkler	___
Painters	___
Tile Setters	___
Carpet Lyrs	___
Controls	___
______	___
______	___
______	___
______	___

SUBCONTRACTORS ON PROJECT & THEIR ACTIVITY

DAILY ACTIVITY - VISITORS - INSPECTIONS

Supervisor's Signature: ______________________________

DAILY LOG

Day: ________________

August 26, 20____ **Weather: 8:00 A.M. ____ Noon: ____ 4:00 P.M. ____**

WORK FORCE

Supt.	____
Foreman	____
Carpenters	____
Laborers	____
Masons	____
Oper, Engrs.	____
Iron Wrkers	____
Electricians	____
Plumbers	____
Steam Fittrs	____
Sheet Metal	____
Glazers	____
Roofers	____
Sprinkler	____
Painters	____
Tile Setters	____
Carpet Lyrs	____
Controls	____
________	____
________	____
________	____
________	____

SUBCONTRACTORS ON PROJECT & THEIR ACTIVITY

DAILY ACTIVITY - VISITORS - INSPECTIONS

Supervisor's Signature: ________________________________

DAILY LOG

Day: ____________________

August 27, 20____ **Weather: 8:00 A.M. ____ Noon: ____ 4:00 P.M. ____**

WORK FORCE	
Supt.	____
Foreman	____
Carpenters	____
Laborers	____
Masons	____
Oper, Engrs.	____
Iron Wrkers	____
Electricians	____
Plumbers	____
Steam Fittrs	____
Sheet Metal	____
Glazers	____
Roofers	____
Sprinkler	____
Painters	____
Tile Setters	____
Carpet Lyrs	____
Controls	____
________	____
________	____
________	____
________	____

SUBCONTRACTORS ON PROJECT & THEIR ACTIVITY

DAILY ACTIVITY - VISITORS - INSPECTIONS

Supervisor's Signature: ______________________________

DAILY LOG

Day: ____________________

August 28, 20___ **Weather: 8:00 A.M. ___ Noon: ___ 4:00 P.M. ___**

WORK FORCE

Supt.	___
Foreman	___
Carpenters	___
Laborers	___
Masons	___
Oper, Engrs.	___
Iron Wrkers	___
Electricians	___
Plumbers	___
Steam Fittrs	___
Sheet Metal	___
Glazers	___
Roofers	___
Sprinkler	___
Painters	___
Tile Setters	___
Carpet Lyrs	___
Controls	___
________	___
________	___
________	___
________	___

SUBCONTRACTORS ON PROJECT & THEIR ACTIVITY

DAILY ACTIVITY - VISITORS - INSPECTIONS

Supervisor's Signature: ______________________________

DAILY LOG

Day: ____________________

August 29, 20____　　**Weather: 8:00 A.M. ___ Noon: ___ 4:00 P.M. ___**

WORK FORCE	
Supt.	____
Foreman	____
Carpenters	____
Laborers	____
Masons	____
Oper, Engrs.	____
Iron Wrkers	____
Electricians	____
Plumbers	____
Steam Fittrs	____
Sheet Metal	____
Glazers	____
Roofers	____
Sprinkler	____
Painters	____
Tile Setters	____
Carpet Lyrs	____
Controls	____
________	____
________	____
________	____
________	____

SUBCONTRACTORS ON PROJECT & THEIR ACTIVITY

DAILY ACTIVITY - VISITORS - INSPECTIONS

Supervisor's Signature: ______________________________

DAILY LOG

Day: ____________________

August 30, 20____ **Weather: 8:00 A.M. ___ Noon: ___ 4:00 P.M. ___**

WORK FORCE	
Supt.	____
Foreman	____
Carpenters	____
Laborers	____
Masons	____
Oper, Engrs.	____
Iron Wrkers	____
Electricians	____
Plumbers	____
Steam Fittrs	____
Sheet Metal	____
Glazers	____
Roofers	____
Sprinkler	____
Painters	____
Tile Setters	____
Carpet Lyrs	____
Controls	____
________	____
________	____
________	____
________	____

SUBCONTRACTORS ON PROJECT & THEIR ACTIVITY

DAILY ACTIVITY - VISITORS - INSPECTIONS

Supervisor's Signature: ______________________________

DAILY LOG

Day: ____________________

August 31, 20____ **Weather: 8:00 A.M. ___ Noon: ___ 4:00 P.M. ___**

WORK FORCE	
Supt.	___
Foreman	___
Carpenters	___
Laborers	___
Masons	___
Oper, Engrs.	___
Iron Wrkers	___
Electricians	___
Plumbers	___
Steam Fittrs	___
Sheet Metal	___
Glazers	___
Roofers	___
Sprinkler	___
Painters	___
Tile Setters	___
Carpet Lyrs	___
Controls	___
________	___
________	___
________	___
________	___

SUBCONTRACTORS ON PROJECT & THEIR ACTIVITY

DAILY ACTIVITY - VISITORS - INSPECTIONS

Supervisor's Signature: ______________________________

DAILY LOG

Day: ____________________

September 1, 20____ **Weather: 8:00 A.M. ___ Noon: ___ 4:00 P.M. ___**

WORK FORCE

Supt.	____
Foreman	____
Carpenters	____
Laborers	____
Masons	____
Oper, Engrs.	____
Iron Wrkers	____
Electricians	____
Plumbers	____
Steam Fittrs	____
Sheet Metal	____
Glazers	____
Roofers	____
Sprinkler	____
Painters	____
Tile Setters	____
Carpet Lyrs	____
Controls	____
________	____
________	____
________	____
________	____

SUBCONTRACTORS ON PROJECT & THEIR ACTIVITY

DAILY ACTIVITY - VISITORS - INSPECTIONS

Supervisor's Signature: ______________________________

DAILY LOG

Day: ____________________

September 2, 20____ **Weather: 8:00 A.M.** ____ **Noon:** ____ **4:00 P.M.** ____

WORK FORCE	
Supt.	____
Foreman	____
Carpenters	____
Laborers	____
Masons	____
Oper, Engrs.	____
Iron Wrkers	____
Electricians	____
Plumbers	____
Steam Fittrs	____
Sheet Metal	____
Glazers	____
Roofers	____
Sprinkler	____
Painters	____
Tile Setters	____
Carpet Lyrs	____
Controls	____
________	____
________	____
________	____
________	____

SUBCONTRACTORS ON PROJECT & THEIR ACTIVITY

DAILY ACTIVITY - VISITORS - INSPECTIONS

Supervisor's Signature: ______________________________

DAILY LOG

Day: ____________________

September 3, 20____ **Weather: 8:00 A.M. ___ Noon: ___ 4:00 P.M. ___**

WORK FORCE

Supt.	___
Foreman	___
Carpenters	___
Laborers	___
Masons	___
Oper, Engrs.	___
Iron Wrkers	___
Electricians	___
Plumbers	___
Steam Fittrs	___
Sheet Metal	___
Glazers	___
Roofers	___
Sprinkler	___
Painters	___
Tile Setters	___
Carpet Lyrs	___
Controls	___
______	___
______	___
______	___
______	___

SUBCONTRACTORS ON PROJECT & THEIR ACTIVITY

DAILY ACTIVITY - VISITORS - INSPECTIONS

Supervisor's Signature: ____________________

DAILY LOG

Day: ____________________

September 4, 20____ **Weather: 8:00 A.M. ___ Noon: ___ 4:00 P.M. ___**

WORK FORCE	
Supt.	___
Foreman	___
Carpenters	___
Laborers	___
Masons	___
Oper, Engrs.	___
Iron Wrkers	___
Electricians	___
Plumbers	___
Steam Fittrs	___
Sheet Metal	___
Glazers	___
Roofers	___
Sprinkler	___
Painters	___
Tile Setters	___
Carpet Lyrs	___
Controls	___
________	___
________	___
________	___
________	___

SUBCONTRACTORS ON PROJECT & THEIR ACTIVITY

DAILY ACTIVITY - VISITORS - INSPECTIONS

Supervisor's Signature: ______________________________

DAILY LOG

Day: ____________________

September 5, 20____ **Weather: 8:00 A.M. ___ Noon: ___ 4:00 P.M. ___**

WORK FORCE

Supt.	___
Foreman	___
Carpenters	___
Laborers	___
Masons	___
Oper, Engrs.	___
Iron Wrkers	___
Electricians	___
Plumbers	___
Steam Fittrs	___
Sheet Metal	___
Glazers	___
Roofers	___
Sprinkler	___
Painters	___
Tile Setters	___
Carpet Lyrs	___
Controls	___
___	___
___	___
___	___
___	___

SUBCONTRACTORS ON PROJECT & THEIR ACTIVITY

DAILY ACTIVITY - VISITORS - INSPECTIONS

Supervisor's Signature: ____________________

DAILY LOG

Day: ____________________

September 6, 20____ **Weather: 8:00 A.M. ___ Noon: ___ 4:00 P.M. ___**

WORK FORCE

Supt.	___
Foreman	___
Carpenters	___
Laborers	___
Masons	___
Oper, Engrs.	___
Iron Wrkers	___
Electricians	___
Plumbers	___
Steam Fittrs	___
Sheet Metal	___
Glazers	___
Roofers	___
Sprinkler	___
Painters	___
Tile Setters	___
Carpet Lyrs	___
Controls	___
______	___
______	___
______	___
______	___

SUBCONTRACTORS ON PROJECT & THEIR ACTIVITY

DAILY ACTIVITY - VISITORS - INSPECTIONS

Supervisor's Signature: ______________________________

DAILY LOG

Day: ____________________

September 7, 20____ **Weather: 8:00 A.M. ___ Noon: ___ 4:00 P.M. ___**

WORK FORCE

Supt.	___
Foreman	___
Carpenters	___
Laborers	___
Masons	___
Oper, Engrs.	___
Iron Wrkers	___
Electricians	___
Plumbers	___
Steam Fittrs	___
Sheet Metal	___
Glazers	___
Roofers	___
Sprinkler	___
Painters	___
Tile Setters	___
Carpet Lyrs	___
Controls	___
______	___
______	___
______	___
______	___

SUBCONTRACTORS ON PROJECT & THEIR ACTIVITY

DAILY ACTIVITY - VISITORS - INSPECTIONS

Supervisor's Signature: ______________________________

DAILY LOG

Day: ____________________

September 8, 20____ **Weather: 8:00 A.M. ____ Noon: ____ 4:00 P.M. ____**

WORK FORCE

Supt.	____
Foreman	____
Carpenters	____
Laborers	____
Masons	____
Oper, Engrs.	____
Iron Wrkers	____
Electricians	____
Plumbers	____
Steam Fittrs	____
Sheet Metal	____
Glazers	____
Roofers	____
Sprinkler	____
Painters	____
Tile Setters	____
Carpet Lyrs	____
Controls	____
________	____
________	____
________	____
________	____

SUBCONTRACTORS ON PROJECT & THEIR ACTIVITY

DAILY ACTIVITY - VISITORS - INSPECTIONS

Supervisor's Signature: ______________________________

DAILY LOG

Day: ____________________

September 9, 20____ **Weather: 8:00 A.M. ___ Noon: ___ 4:00 P.M. ___**

WORK FORCE

Supt.	___
Foreman	___
Carpenters	___
Laborers	___
Masons	___
Oper, Engrs.	___
Iron Wrkers	___
Electricians	___
Plumbers	___
Steam Fittrs	___
Sheet Metal	___
Glazers	___
Roofers	___
Sprinkler	___
Painters	___
Tile Setters	___
Carpet Lyrs	___
Controls	___
______	___
______	___
______	___
______	___

SUBCONTRACTORS ON PROJECT & THEIR ACTIVITY

DAILY ACTIVITY - VISITORS - INSPECTIONS

Supervisor's Signature: ______________________________

DAILY LOG

Day: ____________________

September 10, 20____ **Weather: 8:00 A.M. ___ Noon: ___ 4:00 P.M. ___**

WORK FORCE	
Supt.	___
Foreman	___
Carpenters	___
Laborers	___
Masons	___
Oper, Engrs.	___
Iron Wrkers	___
Electricians	___
Plumbers	___
Steam Fittrs	___
Sheet Metal	___
Glazers	___
Roofers	___
Sprinkler	___
Painters	___
Tile Setters	___
Carpet Lyrs	___
Controls	___
________	___
________	___
________	___
________	___

SUBCONTRACTORS ON PROJECT & THEIR ACTIVITY

DAILY ACTIVITY - VISITORS - INSPECTIONS

Supervisor's Signature: ______________________________

DAILY LOG

Day: ____________________

September 11, 20____ **Weather: 8:00 A.M. ____ Noon: ____ 4:00 P.M. ____**

WORK FORCE	
Supt.	____
Foreman	____
Carpenters	____
Laborers	____
Masons	____
Oper, Engrs.	____
Iron Wrkers	____
Electricians	____
Plumbers	____
Steam Fittrs	____
Sheet Metal	____
Glazers	____
Roofers	____
Sprinkler	____
Painters	____
Tile Setters	____
Carpet Lyrs	____
Controls	____
________	____
________	____
________	____
________	____

SUBCONTRACTORS ON PROJECT & THEIR ACTIVITY

DAILY ACTIVITY - VISITORS - INSPECTIONS

Supervisor's Signature: ______________________________

DAILY LOG

Day: ____________________

September 12, 20____ **Weather: 8:00 A.M. ___ Noon: ___ 4:00 P.M. ___**

WORK FORCE

Supt.	____
Foreman	____
Carpenters	____
Laborers	____
Masons	____
Oper, Engrs.	____
Iron Wrkers	____
Electricians	____
Plumbers	____
Steam Fittrs	____
Sheet Metal	____
Glazers	____
Roofers	____
Sprinkler	____
Painters	____
Tile Setters	____
Carpet Lyrs	____
Controls	____
________	____
________	____
________	____
________	____

SUBCONTRACTORS ON PROJECT & THEIR ACTIVITY

DAILY ACTIVITY - VISITORS - INSPECTIONS

Supervisor's Signature: ______________________________

DAILY LOG

Day: ____________________

September 13, 20____ **Weather: 8:00 A.M. ___ Noon: ___ 4:00 P.M. ___**

WORK FORCE

Supt.	____
Foreman	____
Carpenters	____
Laborers	____
Masons	____
Oper, Engrs.	____
Iron Wrkers	____
Electricians	____
Plumbers	____
Steam Fittrs	____
Sheet Metal	____
Glazers	____
Roofers	____
Sprinkler	____
Painters	____
Tile Setters	____
Carpet Lyrs	____
Controls	____
________	____
________	____
________	____
________	____

SUBCONTRACTORS ON PROJECT & THEIR ACTIVITY

DAILY ACTIVITY - VISITORS - INSPECTIONS

Supervisor's Signature: ______________________________

DAILY LOG

Day: ____________________

September 14, 20____ **Weather: 8:00 A.M. ____ Noon: ____ 4:00 P.M. ____**

WORK FORCE

Supt.	____
Foreman	____
Carpenters	____
Laborers	____
Masons	____
Oper, Engrs.	____
Iron Wrkers	____
Electricians	____
Plumbers	____
Steam Fittrs	____
Sheet Metal	____
Glazers	____
Roofers	____
Sprinkler	____
Painters	____
Tile Setters	____
Carpet Lyrs	____
Controls	____
________	____
________	____
________	____
________	____

SUBCONTRACTORS ON PROJECT & THEIR ACTIVITY

DAILY ACTIVITY - VISITORS - INSPECTIONS

Supervisor's Signature: ____________________

DAILY LOG

Day: ____________________

September 15, 20____ **Weather: 8:00 A.M. ___ Noon: ___ 4:00 P.M. ___**

WORK FORCE	
Supt.	___
Foreman	___
Carpenters	___
Laborers	___
Masons	___
Oper, Engrs.	___
Iron Wrkers	___
Electricians	___
Plumbers	___
Steam Fittrs	___
Sheet Metal	___
Glazers	___
Roofers	___
Sprinkler	___
Painters	___
Tile Setters	___
Carpet Lyrs	___
Controls	___
______	___
______	___
______	___
______	___

SUBCONTRACTORS ON PROJECT & THEIR ACTIVITY

____

DAILY ACTIVITY - VISITORS - INSPECTIONS

____

Supervisor's Signature: ______________________________

DAILY LOG

Day: ____________________

September 16, 20____ **Weather: 8:00 A.M. ___ Noon: ___ 4:00 P.M. ___**

WORK FORCE

Supt.	____
Foreman	____
Carpenters	____
Laborers	____
Masons	____
Oper, Engrs.	____
Iron Wrkers	____
Electricians	____
Plumbers	____
Steam Fittrs	____
Sheet Metal	____
Glazers	____
Roofers	____
Sprinkler	____
Painters	____
Tile Setters	____
Carpet Lyrs	____
Controls	____
________	____
________	____
________	____
________	____

SUBCONTRACTORS ON PROJECT & THEIR ACTIVITY

DAILY ACTIVITY - VISITORS - INSPECTIONS

Supervisor's Signature: ______________________________

DAILY LOG

Day: ____________________

September 17, 20____ **Weather: 8:00 A.M. ___ Noon: ___ 4:00 P.M. ___**

WORK FORCE

Supt.	___
Foreman	___
Carpenters	___
Laborers	___
Masons	___
Oper, Engrs.	___
Iron Wrkers	___
Electricians	___
Plumbers	___
Steam Fittrs	___
Sheet Metal	___
Glazers	___
Roofers	___
Sprinkler	___
Painters	___
Tile Setters	___
Carpet Lyrs	___
Controls	___
________	___
________	___
________	___
________	___

SUBCONTRACTORS ON PROJECT & THEIR ACTIVITY

DAILY ACTIVITY - VISITORS - INSPECTIONS

Supervisor's Signature: ______________________________

DAILY LOG

Day: ____________________

September 18, 20____ **Weather: 8:00 A.M. ___ Noon: ___ 4:00 P.M. ___**

WORK FORCE	
Supt.	___
Foreman	___
Carpenters	___
Laborers	___
Masons	___
Oper, Engrs.	___
Iron Wrkers	___
Electricians	___
Plumbers	___
Steam Fittrs	___
Sheet Metal	___
Glazers	___
Roofers	___
Sprinkler	___
Painters	___
Tile Setters	___
Carpet Lyrs	___
Controls	___
________	___
________	___
________	___
________	___

SUBCONTRACTORS ON PROJECT & THEIR ACTIVITY

DAILY ACTIVITY - VISITORS - INSPECTIONS

Supervisor's Signature: ______________________________

DAILY LOG

Day: ____________________

September 19, 20____ **Weather: 8:00 A.M. ___ Noon: ___ 4:00 P.M. ___**

WORK FORCE

Supt.	___
Foreman	___
Carpenters	___
Laborers	___
Masons	___
Oper, Engrs.	___
Iron Wrkers	___
Electricians	___
Plumbers	___
Steam Fittrs	___
Sheet Metal	___
Glazers	___
Roofers	___
Sprinkler	___
Painters	___
Tile Setters	___
Carpet Lyrs	___
Controls	___
______	___
______	___
______	___
______	___

SUBCONTRACTORS ON PROJECT & THEIR ACTIVITY

DAILY ACTIVITY - VISITORS - INSPECTIONS

Supervisor's Signature: ______________________________

DAILY LOG

Day: ____________________

September 20, 20____ **Weather: 8:00 A.M. ___ Noon: ___ 4:00 P.M. ___**

WORK FORCE

Supt.	___
Foreman	___
Carpenters	___
Laborers	___
Masons	___
Oper, Engrs.	___
Iron Wrkers	___
Electricians	___
Plumbers	___
Steam Fittrs	___
Sheet Metal	___
Glazers	___
Roofers	___
Sprinkler	___
Painters	___
Tile Setters	___
Carpet Lyrs	___
Controls	___
___	___
___	___
___	___
___	___

SUBCONTRACTORS ON PROJECT & THEIR ACTIVITY

DAILY ACTIVITY - VISITORS - INSPECTIONS

Supervisor's Signature: ______________________________

DAILY LOG

Day: ____________________

September 21, 20____ **Weather: 8:00 A.M. ___ Noon: ___ 4:00 P.M. ___**

WORK FORCE

Supt.	___
Foreman	___
Carpenters	___
Laborers	___
Masons	___
Oper, Engrs.	___
Iron Wrkers	___
Electricians	___
Plumbers	___
Steam Fittrs	___
Sheet Metal	___
Glazers	___
Roofers	___
Sprinkler	___
Painters	___
Tile Setters	___
Carpet Lyrs	___
Controls	___
______	___
______	___
______	___
______	___

SUBCONTRACTORS ON PROJECT & THEIR ACTIVITY

DAILY ACTIVITY - VISITORS - INSPECTIONS

Supervisor's Signature: ______________________________

DAILY LOG

Day: ____________________

September 22, 20____ **Weather: 8:00 A.M. ___ Noon: ___ 4:00 P.M. ___**

WORK FORCE

Supt.	___
Foreman	___
Carpenters	___
Laborers	___
Masons	___
Oper, Engrs.	___
Iron Wrkers	___
Electricians	___
Plumbers	___
Steam Fittrs	___
Sheet Metal	___
Glazers	___
Roofers	___
Sprinkler	___
Painters	___
Tile Setters	___
Carpet Lyrs	___
Controls	___
________	___
________	___
________	___
________	___

SUBCONTRACTORS ON PROJECT & THEIR ACTIVITY

DAILY ACTIVITY - VISITORS - INSPECTIONS

Supervisor's Signature: ______________________________

DAILY LOG

Day: ____________________

September 23, 20____ **Weather: 8:00 A.M. ____ Noon: ____ 4:00 P.M. ____**

WORK FORCE **SUBCONTRACTORS ON PROJECT & THEIR ACTIVITY**

WORK FORCE		SUBCONTRACTORS ON PROJECT & THEIR ACTIVITY
Supt.	____	____________________
Foreman	____	____________________
Carpenters	____	____________________
Laborers	____	____________________
Masons	____	____________________
Oper, Engrs.	____	____________________
Iron Wrkers	____	____________________
Electricians	____	____________________
Plumbers	____	____________________
Steam Fittrs	____	____________________
Sheet Metal	____	____________________
Glazers	____	____________________
Roofers	____	____________________
Sprinkler	____	____________________
Painters	____	____________________
Tile Setters	____	____________________
Carpet Lyrs	____	____________________
Controls	____	____________________
____	____	DAILY ACTIVITY - VISITORS - INSPECTIONS
____	____	____________________
____	____	____________________
____	____	____________________
		____________________

____

____

____

Supervisor's Signature: ____________________

DAILY LOG

Day: ____________________

September 24, 20____ **Weather: 8:00 A.M. ___ Noon: ___ 4:00 P.M. ___**

WORK FORCE

Supt.	___
Foreman	___
Carpenters	___
Laborers	___
Masons	___
Oper, Engrs.	___
Iron Wrkers	___
Electricians	___
Plumbers	___
Steam Fittrs	___
Sheet Metal	___
Glazers	___
Roofers	___
Sprinkler	___
Painters	___
Tile Setters	___
Carpet Lyrs	___
Controls	___
______	___
______	___
______	___
______	___

SUBCONTRACTORS ON PROJECT & THEIR ACTIVITY

DAILY ACTIVITY - VISITORS - INSPECTIONS

Supervisor's Signature: ______________________________

DAILY LOG

Day: ____________________

September 25, 20____ **Weather: 8:00 A.M. ___ Noon: ___ 4:00 P.M. ___**

WORK FORCE		SUBCONTRACTORS ON PROJECT & THEIR ACTIVITY
Supt.	___	
Foreman	___	
Carpenters	___	
Laborers	___	
Masons	___	
Oper, Engrs.	___	
Iron Wrkers	___	
Electricians	___	
Plumbers	___	
Steam Fittrs	___	
Sheet Metal	___	
Glazers	___	
Roofers	___	
Sprinkler	___	
Painters	___	
Tile Setters	___	
Carpet Lyrs	___	
Controls	___	
___	___	DAILY ACTIVITY - VISITORS - INSPECTIONS
___	___	
___	___	
___	___	

Supervisor's Signature: ____________________

DAILY LOG

Day: ____________________

September 26, 20____ **Weather: 8:00 A.M. ___ Noon: ___ 4:00 P.M. ___**

WORK FORCE

Supt.	___
Foreman	___
Carpenters	___
Laborers	___
Masons	___
Oper, Engrs.	___
Iron Wrkers	___
Electricians	___
Plumbers	___
Steam Fittrs	___
Sheet Metal	___
Glazers	___
Roofers	___
Sprinkler	___
Painters	___
Tile Setters	___
Carpet Lyrs	___
Controls	___
______	___
______	___
______	___
______	___

SUBCONTRACTORS ON PROJECT & THEIR ACTIVITY

DAILY ACTIVITY - VISITORS - INSPECTIONS

Supervisor's Signature: ______________________________

DAILY LOG

Day: ____________________

September 27, 20____ **Weather: 8:00 A.M. ___ Noon: ___ 4:00 P.M. ___**

WORK FORCE **SUBCONTRACTORS ON PROJECT & THEIR ACTIVITY**

WORK FORCE	
Supt.	___
Foreman	___
Carpenters	___
Laborers	___
Masons	___
Oper, Engrs.	___
Iron Wrkers	___
Electricians	___
Plumbers	___
Steam Fittrs	___
Sheet Metal	___
Glazers	___
Roofers	___
Sprinkler	___
Painters	___
Tile Setters	___
Carpet Lyrs	___
Controls	___
___	___
___	___
___	___
___	___

DAILY ACTIVITY - VISITORS - INSPECTIONS

Supervisor's Signature: ____________________

DAILY LOG

Day: ____________________

September 28, 20____ **Weather: 8:00 A.M. ___ Noon: ___ 4:00 P.M. ___**

WORK FORCE		SUBCONTRACTORS ON PROJECT & THEIR ACTIVITY
Supt.	____	
Foreman	____	
Carpenters	____	
Laborers	____	
Masons	____	
Oper, Engrs.	____	
Iron Wrkers	____	
Electricians	____	
Plumbers	____	
Steam Fittrs	____	
Sheet Metal	____	
Glazers	____	
Roofers	____	
Sprinkler	____	
Painters	____	
Tile Setters	____	
Carpet Lyrs	____	
Controls	____	
________	____	DAILY ACTIVITY - VISITORS - INSPECTIONS
________	____	
________	____	
________	____	

Supervisor's Signature: ________________________________

DAILY LOG

Day: ____________________

September 29, 20____ **Weather: 8:00 A.M. ____ Noon: ____ 4:00 P.M. ____**

WORK FORCE

Supt.	____
Foreman	____
Carpenters	____
Laborers	____
Masons	____
Oper, Engrs.	____
Iron Wrkers	____
Electricians	____
Plumbers	____
Steam Fittrs	____
Sheet Metal	____
Glazers	____
Roofers	____
Sprinkler	____
Painters	____
Tile Setters	____
Carpet Lyrs	____
Controls	____
________	____
________	____
________	____
________	____

SUBCONTRACTORS ON PROJECT & THEIR ACTIVITY

DAILY ACTIVITY - VISITORS - INSPECTIONS

Supervisor's Signature: ______________________________

DAILY LOG

Day: ____________

September 30, 20____ **Weather: 8:00 A.M. ____ Noon: ____ 4:00 P.M. ____**

WORK FORCE	
Supt.	____
Foreman	____
Carpenters	____
Laborers	____
Masons	____
Oper, Engrs.	____
Iron Wrkers	____
Electricians	____
Plumbers	____
Steam Fittrs	____
Sheet Metal	____
Glazers	____
Roofers	____
Sprinkler	____
Painters	____
Tile Setters	____
Carpet Lyrs	____
Controls	____
____	____
____	____
____	____
____	____

SUBCONTRACTORS ON PROJECT & THEIR ACTIVITY

DAILY ACTIVITY - VISITORS - INSPECTIONS

Supervisor's Signature: ____________

DAILY LOG

Day: ____________________

October 1, 20____ **Weather: 8:00 A.M.** ___ **Noon:** ___ **4:00 P.M.** ___

WORK FORCE	
Supt.	___
Foreman	___
Carpenters	___
Laborers	___
Masons	___
Oper, Engrs.	___
Iron Wrkers	___
Electricians	___
Plumbers	___
Steam Fittrs	___
Sheet Metal	___
Glazers	___
Roofers	___
Sprinkler	___
Painters	___
Tile Setters	___
Carpet Lyrs	___
Controls	___
______	___
______	___
______	___
______	___

SUBCONTRACTORS ON PROJECT & THEIR ACTIVITY

DAILY ACTIVITY - VISITORS - INSPECTIONS

Supervisor's Signature: ______________________________

DAILY LOG

Day: ________________

October 2, 20___ **Weather: 8:00 A.M. ___ Noon: ___ 4:00 P.M. ___**

WORK FORCE

Supt.	___
Foreman	___
Carpenters	___
Laborers	___
Masons	___
Oper, Engrs.	___
Iron Wrkers	___
Electricians	___
Plumbers	___
Steam Fittrs	___
Sheet Metal	___
Glazers	___
Roofers	___
Sprinkler	___
Painters	___
Tile Setters	___
Carpet Lyrs	___
Controls	___
________	___
________	___
________	___
________	___

SUBCONTRACTORS ON PROJECT & THEIR ACTIVITY

DAILY ACTIVITY - VISITORS - INSPECTIONS

Supervisor's Signature: ______________________________

DAILY LOG

Day: ____________________

October 3, 20____ **Weather: 8:00 A.M. ____ Noon: ____ 4:00 P.M. ____**

WORK FORCE | **SUBCONTRACTORS ON PROJECT & THEIR ACTIVITY**

WORK FORCE	
Supt.	____
Foreman	____
Carpenters	____
Laborers	____
Masons	____
Oper, Engrs.	____
Iron Wrkers	____
Electricians	____
Plumbers	____
Steam Fittrs	____
Sheet Metal	____
Glazers	____
Roofers	____
Sprinkler	____
Painters	____
Tile Setters	____
Carpet Lyrs	____
Controls	____
________	____
________	____
________	____
________	____

DAILY ACTIVITY - VISITORS - INSPECTIONS

Supervisor's Signature: ________________________________

DAILY LOG

Day: ____________________

October 4, 20____ **Weather: 8:00 A.M. ___ Noon: ___ 4:00 P.M. ___**

WORK FORCE	
Supt.	___
Foreman	___
Carpenters	___
Laborers	___
Masons	___
Oper, Engrs.	___
Iron Wrkers	___
Electricians	___
Plumbers	___
Steam Fittrs	___
Sheet Metal	___
Glazers	___
Roofers	___
Sprinkler	___
Painters	___
Tile Setters	___
Carpet Lyrs	___
Controls	___
________	___
________	___
________	___
________	___

SUBCONTRACTORS ON PROJECT & THEIR ACTIVITY

DAILY ACTIVITY - VISITORS - INSPECTIONS

Supervisor's Signature: ______________________________

DAILY LOG

Day: ____________________

October 5, 20____ **Weather: 8:00 A.M. ___ Noon: ___ 4:00 P.M. ___**

WORK FORCE	
Supt.	____
Foreman	____
Carpenters	____
Laborers	____
Masons	____
Oper, Engrs.	____
Iron Wrkers	____
Electricians	____
Plumbers	____
Steam Fittrs	____
Sheet Metal	____
Glazers	____
Roofers	____
Sprinkler	____
Painters	____
Tile Setters	____
Carpet Lyrs	____
Controls	____
________	____
________	____
________	____
________	____

SUBCONTRACTORS ON PROJECT & THEIR ACTIVITY

DAILY ACTIVITY - VISITORS - INSPECTIONS

Supervisor's Signature: ______________________________

DAILY LOG

Day: ____________________

October 6, 20____ **Weather: 8:00 A.M. ___ Noon: ___ 4:00 P.M. ___**

WORK FORCE	
Supt.	___
Foreman	___
Carpenters	___
Laborers	___
Masons	___
Oper, Engrs.	___
Iron Wrkers	___
Electricians	___
Plumbers	___
Steam Fittrs	___
Sheet Metal	___
Glazers	___
Roofers	___
Sprinkler	___
Painters	___
Tile Setters	___
Carpet Lyrs	___
Controls	___
________	___
________	___
________	___
________	___

SUBCONTRACTORS ON PROJECT & THEIR ACTIVITY

DAILY ACTIVITY - VISITORS - INSPECTIONS

Supervisor's Signature: ______________________________

DAILY LOG

Day: ____________________

October 7, 20____ **Weather: 8:00 A.M. ____ Noon: ____ 4:00 P.M. ____**

WORK FORCE

Supt.	____
Foreman	____
Carpenters	____
Laborers	____
Masons	____
Oper, Engrs.	____
Iron Wrkers	____
Electricians	____
Plumbers	____
Steam Fittrs	____
Sheet Metal	____
Glazers	____
Roofers	____
Sprinkler	____
Painters	____
Tile Setters	____
Carpet Lyrs	____
Controls	____
________	____
________	____
________	____
________	____

SUBCONTRACTORS ON PROJECT & THEIR ACTIVITY

DAILY ACTIVITY - VISITORS - INSPECTIONS

Supervisor's Signature: ________________________________

DAILY LOG

Day: ____________________

October 8, 20____　　**Weather: 8:00 A.M. ___ Noon: ___ 4:00 P.M. ___**

WORK FORCE	
Supt.	___
Foreman	___
Carpenters	___
Laborers	___
Masons	___
Oper, Engrs.	___
Iron Wrkers	___
Electricians	___
Plumbers	___
Steam Fittrs	___
Sheet Metal	___
Glazers	___
Roofers	___
Sprinkler	___
Painters	___
Tile Setters	___
Carpet Lyrs	___
Controls	___
______	___
______	___
______	___
______	___

SUBCONTRACTORS ON PROJECT & THEIR ACTIVITY

DAILY ACTIVITY - VISITORS - INSPECTIONS

Supervisor's Signature: ______________________________

DAILY LOG

Day: ____________________

October 9, 20____ **Weather: 8:00 A.M. ___ Noon: ___ 4:00 P.M. ___**

WORK FORCE

Supt.	____
Foreman	____
Carpenters	____
Laborers	____
Masons	____
Oper, Engrs.	____
Iron Wrkers	____
Electricians	____
Plumbers	____
Steam Fittrs	____
Sheet Metal	____
Glazers	____
Roofers	____
Sprinkler	____
Painters	____
Tile Setters	____
Carpet Lyrs	____
Controls	____
________	____
________	____
________	____
________	____

SUBCONTRACTORS ON PROJECT & THEIR ACTIVITY

DAILY ACTIVITY - VISITORS - INSPECTIONS

Supervisor's Signature: ______________________________

DAILY LOG

Day: ____________________

October 10, 20____ **Weather: 8:00 A.M. ___ Noon: ___ 4:00 P.M. ___**

WORK FORCE

Supt.	___
Foreman	___
Carpenters	___
Laborers	___
Masons	___
Oper, Engrs.	___
Iron Wrkers	___
Electricians	___
Plumbers	___
Steam Fittrs	___
Sheet Metal	___
Glazers	___
Roofers	___
Sprinkler	___
Painters	___
Tile Setters	___
Carpet Lyrs	___
Controls	___
______	___
______	___
______	___
______	___

SUBCONTRACTORS ON PROJECT & THEIR ACTIVITY

DAILY ACTIVITY - VISITORS - INSPECTIONS

Supervisor's Signature: ______________________________

DAILY LOG

Day: ____________________

October 11, 20____ **Weather: 8:00 A.M.** ___ **Noon:** ___ **4:00 P.M.** ___

WORK FORCE

Supt.	___
Foreman	___
Carpenters	___
Laborers	___
Masons	___
Oper, Engrs.	___
Iron Wrkers	___
Electricians	___
Plumbers	___
Steam Fittrs	___
Sheet Metal	___
Glazers	___
Roofers	___
Sprinkler	___
Painters	___
Tile Setters	___
Carpet Lyrs	___
Controls	___
___	___
___	___
___	___
___	___

SUBCONTRACTORS ON PROJECT & THEIR ACTIVITY

DAILY ACTIVITY - VISITORS - INSPECTIONS

Supervisor's Signature: ______________________________

DAILY LOG

Day: ____________________

October 12, 20____ **Weather: 8:00 A.M. ___ Noon: ___ 4:00 P.M. ___**

WORK FORCE

Supt.	___
Foreman	___
Carpenters	___
Laborers	___
Masons	___
Oper, Engrs.	___
Iron Wrkers	___
Electricians	___
Plumbers	___
Steam Fittrs	___
Sheet Metal	___
Glazers	___
Roofers	___
Sprinkler	___
Painters	___
Tile Setters	___
Carpet Lyrs	___
Controls	___
________	___
________	___
________	___
________	___

SUBCONTRACTORS ON PROJECT & THEIR ACTIVITY

DAILY ACTIVITY - VISITORS - INSPECTIONS

Supervisor's Signature: ________________________

DAILY LOG

Day: ____________________

October 13, 20____ **Weather: 8:00 A.M. ___ Noon: ___ 4:00 P.M. ___**

WORK FORCE	
Supt.	___
Foreman	___
Carpenters	___
Laborers	___
Masons	___
Oper, Engrs.	___
Iron Wrkers	___
Electricians	___
Plumbers	___
Steam Fittrs	___
Sheet Metal	___
Glazers	___
Roofers	___
Sprinkler	___
Painters	___
Tile Setters	___
Carpet Lyrs	___
Controls	___
______	___
______	___
______	___
______	___

SUBCONTRACTORS ON PROJECT & THEIR ACTIVITY

DAILY ACTIVITY - VISITORS - INSPECTIONS

Supervisor's Signature: ______________________________

DAILY LOG

Day: ____________________

October 14, 20____ **Weather: 8:00 A.M. ____ Noon: ____ 4:00 P.M. ____**

WORK FORCE

Supt.	____
Foreman	____
Carpenters	____
Laborers	____
Masons	____
Oper, Engrs.	____
Iron Wrkers	____
Electricians	____
Plumbers	____
Steam Fittrs	____
Sheet Metal	____
Glazers	____
Roofers	____
Sprinkler	____
Painters	____
Tile Setters	____
Carpet Lyrs	____
Controls	____
________	____
________	____
________	____
________	____

SUBCONTRACTORS ON PROJECT & THEIR ACTIVITY

DAILY ACTIVITY - VISITORS - INSPECTIONS

Supervisor's Signature: ______________________________

DAILY LOG

Day: ____________________

October 15, 20____ **Weather: 8:00 A.M.** ____ **Noon:** ____ **4:00 P.M.** ____

WORK FORCE

Supt.	____
Foreman	____
Carpenters	____
Laborers	____
Masons	____
Oper, Engrs.	____
Iron Wrkers	____
Electricians	____
Plumbers	____
Steam Fittrs	____
Sheet Metal	____
Glazers	____
Roofers	____
Sprinkler	____
Painters	____
Tile Setters	____
Carpet Lyrs	____
Controls	____
________	____
________	____
________	____
________	____

SUBCONTRACTORS ON PROJECT & THEIR ACTIVITY

DAILY ACTIVITY - VISITORS - INSPECTIONS

Supervisor's Signature: ______________________________

DAILY LOG

Day: ____________________

October 16, 20____ **Weather: 8:00 A.M. ___ Noon: ___ 4:00 P.M. ___**

WORK FORCE	
Supt.	___
Foreman	___
Carpenters	___
Laborers	___
Masons	___
Oper, Engrs.	___
Iron Wrkers	___
Electricians	___
Plumbers	___
Steam Fittrs	___
Sheet Metal	___
Glazers	___
Roofers	___
Sprinkler	___
Painters	___
Tile Setters	___
Carpet Lyrs	___
Controls	___
________	___
________	___
________	___
________	___

SUBCONTRACTORS ON PROJECT & THEIR ACTIVITY

DAILY ACTIVITY - VISITORS - INSPECTIONS

Supervisor's Signature: ______________________________

DAILY LOG

Day: ____________________

October 17, 20____ **Weather: 8:00 A.M. ___ Noon: ___ 4:00 P.M. ___**

WORK FORCE

Supt.	___
Foreman	___
Carpenters	___
Laborers	___
Masons	___
Oper, Engrs.	___
Iron Wrkers	___
Electricians	___
Plumbers	___
Steam Fittrs	___
Sheet Metal	___
Glazers	___
Roofers	___
Sprinkler	___
Painters	___
Tile Setters	___
Carpet Lyrs	___
Controls	___
______	___
______	___
______	___
______	___

SUBCONTRACTORS ON PROJECT & THEIR ACTIVITY

DAILY ACTIVITY - VISITORS - INSPECTIONS

Supervisor's Signature: ______________________________

DAILY LOG

Day: ____________________

October 18, 20____ **Weather: 8:00 A.M. ___ Noon: ___ 4:00 P.M. ___**

WORK FORCE	
Supt.	___
Foreman	___
Carpenters	___
Laborers	___
Masons	___
Oper, Engrs.	___
Iron Wrkers	___
Electricians	___
Plumbers	___
Steam Fittrs	___
Sheet Metal	___
Glazers	___
Roofers	___
Sprinkler	___
Painters	___
Tile Setters	___
Carpet Lyrs	___
Controls	___
________	___
________	___
________	___
________	___

SUBCONTRACTORS ON PROJECT & THEIR ACTIVITY

DAILY ACTIVITY - VISITORS - INSPECTIONS

Supervisor's Signature: ____________________

DAILY LOG

Day: ____________________

October 19, 20____ **Weather: 8:00 A.M. ____ Noon: ____ 4:00 P.M. ____**

WORK FORCE	
Supt.	____
Foreman	____
Carpenters	____
Laborers	____
Masons	____
Oper, Engrs.	____
Iron Wrkers	____
Electricians	____
Plumbers	____
Steam Fittrs	____
Sheet Metal	____
Glazers	____
Roofers	____
Sprinkler	____
Painters	____
Tile Setters	____
Carpet Lyrs	____
Controls	____
________	____
________	____
________	____
________	____

SUBCONTRACTORS ON PROJECT & THEIR ACTIVITY

DAILY ACTIVITY - VISITORS - INSPECTIONS

Supervisor's Signature: ______________________________

DAILY LOG

Day: ____________________

October 20, 20____ **Weather: 8:00 A.M. ___ Noon: ___ 4:00 P.M. ___**

WORK FORCE	
Supt.	___
Foreman	___
Carpenters	___
Laborers	___
Masons	___
Oper, Engrs.	___
Iron Wrkers	___
Electricians	___
Plumbers	___
Steam Fittrs	___
Sheet Metal	___
Glazers	___
Roofers	___
Sprinkler	___
Painters	___
Tile Setters	___
Carpet Lyrs	___
Controls	___
________	___
________	___
________	___
________	___

SUBCONTRACTORS ON PROJECT & THEIR ACTIVITY

DAILY ACTIVITY - VISITORS - INSPECTIONS

Supervisor's Signature: ______________________________

DAILY LOG

Day: ____________________

October 21, 20____ **Weather: 8:00 A.M. ____ Noon: ____ 4:00 P.M. ____**

WORK FORCE

Supt.	____
Foreman	____
Carpenters	____
Laborers	____
Masons	____
Oper, Engrs.	____
Iron Wrkers	____
Electricians	____
Plumbers	____
Steam Fittrs	____
Sheet Metal	____
Glazers	____
Roofers	____
Sprinkler	____
Painters	____
Tile Setters	____
Carpet Lyrs	____
Controls	____
________	____
________	____
________	____
________	____

SUBCONTRACTORS ON PROJECT & THEIR ACTIVITY

DAILY ACTIVITY - VISITORS - INSPECTIONS

Supervisor's Signature: ______________________________

DAILY LOG

Day: ____________________

October 22, 20____ **Weather: 8:00 A.M. ___ Noon: ___ 4:00 P.M. ___**

WORK FORCE	
Supt.	___
Foreman	___
Carpenters	___
Laborers	___
Masons	___
Oper, Engrs.	___
Iron Wrkers	___
Electricians	___
Plumbers	___
Steam Fittrs	___
Sheet Metal	___
Glazers	___
Roofers	___
Sprinkler	___
Painters	___
Tile Setters	___
Carpet Lyrs	___
Controls	___
______	___
______	___
______	___
______	___

SUBCONTRACTORS ON PROJECT & THEIR ACTIVITY

DAILY ACTIVITY - VISITORS - INSPECTIONS

Supervisor's Signature: ______________________________

DAILY LOG

Day: ____________________

October 23, 20____ **Weather: 8:00 A.M. ____ Noon: ____ 4:00 P.M. ____**

WORK FORCE

Supt.	____
Foreman	____
Carpenters	____
Laborers	____
Masons	____
Oper, Engrs.	____
Iron Wrkers	____
Electricians	____
Plumbers	____
Steam Fittrs	____
Sheet Metal	____
Glazers	____
Roofers	____
Sprinkler	____
Painters	____
Tile Setters	____
Carpet Lyrs	____
Controls	____
________	____
________	____
________	____
________	____

SUBCONTRACTORS ON PROJECT & THEIR ACTIVITY

DAILY ACTIVITY - VISITORS - INSPECTIONS

Supervisor's Signature: ______________________________

DAILY LOG

Day: ____________________

October 24, 20____ **Weather: 8:00 A.M. ___ Noon: ___ 4:00 P.M. ___**

WORK FORCE

Supt.	___
Foreman	___
Carpenters	___
Laborers	___
Masons	___
Oper, Engrs.	___
Iron Wrkers	___
Electricians	___
Plumbers	___
Steam Fittrs	___
Sheet Metal	___
Glazers	___
Roofers	___
Sprinkler	___
Painters	___
Tile Setters	___
Carpet Lyrs	___
Controls	___
________	___
________	___
________	___
________	___

SUBCONTRACTORS ON PROJECT & THEIR ACTIVITY

DAILY ACTIVITY - VISITORS - INSPECTIONS

Supervisor's Signature: ____________________

DAILY LOG

Day: ____________________

October 25, 20____ **Weather: 8:00 A.M. ___ Noon: ___ 4:00 P.M. ___**

WORK FORCE

Supt.	___
Foreman	___
Carpenters	___
Laborers	___
Masons	___
Oper, Engrs.	___
Iron Wrkers	___
Electricians	___
Plumbers	___
Steam Fittrs	___
Sheet Metal	___
Glazers	___
Roofers	___
Sprinkler	___
Painters	___
Tile Setters	___
Carpet Lyrs	___
Controls	___
______	___
______	___
______	___
______	___

SUBCONTRACTORS ON PROJECT & THEIR ACTIVITY

DAILY ACTIVITY - VISITORS - INSPECTIONS

Supervisor's Signature: ______________________________

DAILY LOG

Day: ____________________

October 26, 20____ **Weather: 8:00 A.M.** ____ **Noon:** ____ **4:00 P.M.** ____

WORK FORCE

Supt.	____
Foreman	____
Carpenters	____
Laborers	____
Masons	____
Oper, Engrs.	____
Iron Wrkers	____
Electricians	____
Plumbers	____
Steam Fittrs	____
Sheet Metal	____
Glazers	____
Roofers	____
Sprinkler	____
Painters	____
Tile Setters	____
Carpet Lyrs	____
Controls	____
________	____
________	____
________	____
________	____

SUBCONTRACTORS ON PROJECT & THEIR ACTIVITY

DAILY ACTIVITY - VISITORS - INSPECTIONS

Supervisor's Signature: ______________________________

DAILY LOG

Day: ____________________

October 27, 20____ **Weather: 8:00 A.M. ____ Noon: ____ 4:00 P.M. ____**

WORK FORCE

Supt.	____
Foreman	____
Carpenters	____
Laborers	____
Masons	____
Oper, Engrs.	____
Iron Wrkers	____
Electricians	____
Plumbers	____
Steam Fittrs	____
Sheet Metal	____
Glazers	____
Roofers	____
Sprinkler	____
Painters	____
Tile Setters	____
Carpet Lyrs	____
Controls	____
________	____
________	____
________	____
________	____

SUBCONTRACTORS ON PROJECT & THEIR ACTIVITY

DAILY ACTIVITY - VISITORS - INSPECTIONS

Supervisor's Signature: ________________________________

DAILY LOG

Day: ____________________

October 28, 20____ **Weather: 8:00 A.M. ___ Noon: ___ 4:00 P.M. ___**

WORK FORCE	
Supt.	___
Foreman	___
Carpenters	___
Laborers	___
Masons	___
Oper, Engrs.	___
Iron Wrkers	___
Electricians	___
Plumbers	___
Steam Fittrs	___
Sheet Metal	___
Glazers	___
Roofers	___
Sprinkler	___
Painters	___
Tile Setters	___
Carpet Lyrs	___
Controls	___
________	___
________	___
________	___
________	___

SUBCONTRACTORS ON PROJECT & THEIR ACTIVITY

DAILY ACTIVITY - VISITORS - INSPECTIONS

Supervisor's Signature: ______________________________

DAILY LOG

Day: ____________________

October 29, 20____ **Weather: 8:00 A.M. ___ Noon: ___ 4:00 P.M. ___**

WORK FORCE

Supt.	___
Foreman	___
Carpenters	___
Laborers	___
Masons	___
Oper, Engrs.	___
Iron Wrkers	___
Electricians	___
Plumbers	___
Steam Fittrs	___
Sheet Metal	___
Glazers	___
Roofers	___
Sprinkler	___
Painters	___
Tile Setters	___
Carpet Lyrs	___
Controls	___
______	___
______	___
______	___
______	___

SUBCONTRACTORS ON PROJECT & THEIR ACTIVITY

DAILY ACTIVITY - VISITORS - INSPECTIONS

Supervisor's Signature: ____________________

DAILY LOG

Day: ______________________

October 30, 20____ **Weather: 8:00 A.M. ____ Noon: ____ 4:00 P.M. ____**

WORK FORCE | **SUBCONTRACTORS ON PROJECT & THEIR ACTIVITY**

Work Force	
Supt.	____
Foreman	____
Carpenters	____
Laborers	____
Masons	____
Oper, Engrs.	____
Iron Wrkers	____
Electricians	____
Plumbers	____
Steam Fittrs	____
Sheet Metal	____
Glazers	____
Roofers	____
Sprinkler	____
Painters	____
Tile Setters	____
Carpet Lyrs	____
Controls	____
________	____
________	____
________	____
________	____

DAILY ACTIVITY - VISITORS - INSPECTIONS

Supervisor's Signature: ______________________________

DAILY LOG

Day: ____________________

October 31, 20____ **Weather: 8:00 A.M. ___ Noon: ___ 4:00 P.M. ___**

WORK FORCE		**SUBCONTRACTORS ON PROJECT & THEIR ACTIVITY**
Supt.	___	
Foreman	___	
Carpenters	___	
Laborers	___	
Masons	___	
Oper, Engrs.	___	
Iron Wrkers	___	
Electricians	___	
Plumbers	___	
Steam Fittrs	___	
Sheet Metal	___	
Glazers	___	
Roofers	___	
Sprinkler	___	
Painters	___	
Tile Setters	___	
Carpet Lyrs	___	
Controls	___	
___	___	DAILY ACTIVITY - VISITORS - INSPECTIONS
___	___	
___	___	
___	___	

Supervisor's Signature: ______________________________

DAILY LOG

Day: ____________________

November 1, 20____ **Weather: 8:00 A.M. ___ Noon: ___ 4:00 P.M. ___**

WORK FORCE

Supt.	___
Foreman	___
Carpenters	___
Laborers	___
Masons	___
Oper, Engrs.	___
Iron Wrkers	___
Electricians	___
Plumbers	___
Steam Fittrs	___
Sheet Metal	___
Glazers	___
Roofers	___
Sprinkler	___
Painters	___
Tile Setters	___
Carpet Lyrs	___
Controls	___
________	___
________	___
________	___
________	___

SUBCONTRACTORS ON PROJECT & THEIR ACTIVITY

DAILY ACTIVITY - VISITORS - INSPECTIONS

Supervisor's Signature: ____________________

DAILY LOG

Day: ____________________

November 2, 20____ **Weather: 8:00 A.M. ___ Noon: ___ 4:00 P.M. ___**

WORK FORCE

Supt.	___
Foreman	___
Carpenters	___
Laborers	___
Masons	___
Oper, Engrs.	___
Iron Wrkers	___
Electricians	___
Plumbers	___
Steam Fittrs	___
Sheet Metal	___
Glazers	___
Roofers	___
Sprinkler	___
Painters	___
Tile Setters	___
Carpet Lyrs	___
Controls	___
______	___
______	___
______	___
______	___

SUBCONTRACTORS ON PROJECT & THEIR ACTIVITY

DAILY ACTIVITY - VISITORS - INSPECTIONS

Supervisor's Signature: ______________________________

DAILY LOG

Day: ____________________

November 3, 20____ **Weather: 8:00 A.M. ___ Noon: ___ 4:00 P.M. ___**

WORK FORCE

Supt.	___
Foreman	___
Carpenters	___
Laborers	___
Masons	___
Oper, Engrs.	___
Iron Wrkers	___
Electricians	___
Plumbers	___
Steam Fittrs	___
Sheet Metal	___
Glazers	___
Roofers	___
Sprinkler	___
Painters	___
Tile Setters	___
Carpet Lyrs	___
Controls	___
______	___
______	___
______	___
______	___

SUBCONTRACTORS ON PROJECT & THEIR ACTIVITY

DAILY ACTIVITY - VISITORS - INSPECTIONS

Supervisor's Signature: ______________________________

DAILY LOG

Day: ______________

November 4, 20___ **Weather: 8:00 A.M. ___ Noon: ___ 4:00 P.M. ___**

WORK FORCE

Supt.	___
Foreman	___
Carpenters	___
Laborers	___
Masons	___
Oper, Engrs.	___
Iron Wrkers	___
Electricians	___
Plumbers	___
Steam Fittrs	___
Sheet Metal	___
Glazers	___
Roofers	___
Sprinkler	___
Painters	___
Tile Setters	___
Carpet Lyrs	___
Controls	___
___	___
___	___
___	___
___	___

SUBCONTRACTORS ON PROJECT & THEIR ACTIVITY

DAILY ACTIVITY - VISITORS - INSPECTIONS

Supervisor's Signature: ______________

DAILY LOG

Day: ____________________

November 5, 20____ **Weather: 8:00 A.M. ___ Noon: ___ 4:00 P.M. ___**

WORK FORCE	
Supt.	___
Foreman	___
Carpenters	___
Laborers	___
Masons	___
Oper, Engrs.	___
Iron Wrkers	___
Electricians	___
Plumbers	___
Steam Fittrs	___
Sheet Metal	___
Glazers	___
Roofers	___
Sprinkler	___
Painters	___
Tile Setters	___
Carpet Lyrs	___
Controls	___
________	___
________	___
________	___
________	___

SUBCONTRACTORS ON PROJECT & THEIR ACTIVITY

____

DAILY ACTIVITY - VISITORS - INSPECTIONS

____

Supervisor's Signature: ______________________________

DAILY LOG

Day: ____________________

November 6, 20____ **Weather: 8:00 A.M. ____ Noon: ____ 4:00 P.M. ____**

WORK FORCE	
Supt.	____
Foreman	____
Carpenters	____
Laborers	____
Masons	____
Oper, Engrs.	____
Iron Wrkers	____
Electricians	____
Plumbers	____
Steam Fittrs	____
Sheet Metal	____
Glazers	____
Roofers	____
Sprinkler	____
Painters	____
Tile Setters	____
Carpet Lyrs	____
Controls	____
________	____
________	____
________	____
________	____

SUBCONTRACTORS ON PROJECT & THEIR ACTIVITY

DAILY ACTIVITY - VISITORS - INSPECTIONS

Supervisor's Signature: ______________________________

DAILY LOG

Day: ________________

November 7, 20____ **Weather: 8:00 A.M. ____ Noon: ____ 4:00 P.M. ____**

WORK FORCE	
Supt.	____
Foreman	____
Carpenters	____
Laborers	____
Masons	____
Oper, Engrs.	____
Iron Wrkers	____
Electricians	____
Plumbers	____
Steam Fittrs	____
Sheet Metal	____
Glazers	____
Roofers	____
Sprinkler	____
Painters	____
Tile Setters	____
Carpet Lyrs	____
Controls	____
________	____
________	____
________	____
________	____

SUBCONTRACTORS ON PROJECT & THEIR ACTIVITY

DAILY ACTIVITY - VISITORS - INSPECTIONS

Supervisor's Signature: ____________________

DAILY LOG

Day: ____________________

November 8, 20____ **Weather: 8:00 A.M. ___ Noon: ___ 4:00 P.M. ___**

WORK FORCE	
Supt.	___
Foreman	___
Carpenters	___
Laborers	___
Masons	___
Oper, Engrs.	___
Iron Wrkers	___
Electricians	___
Plumbers	___
Steam Fittrs	___
Sheet Metal	___
Glazers	___
Roofers	___
Sprinkler	___
Painters	___
Tile Setters	___
Carpet Lyrs	___
Controls	___
___	___
___	___
___	___
___	___

SUBCONTRACTORS ON PROJECT & THEIR ACTIVITY

DAILY ACTIVITY - VISITORS - INSPECTIONS

Supervisor's Signature: ____________________

DAILY LOG

Day: ____________________

November 9, 20____   **Weather: 8:00 A.M. ____ Noon: ____ 4:00 P.M. ____**

WORK FORCE	
Supt.	____
Foreman	____
Carpenters	____
Laborers	____
Masons	____
Oper, Engrs.	____
Iron Wrkers	____
Electricians	____
Plumbers	____
Steam Fittrs	____
Sheet Metal	____
Glazers	____
Roofers	____
Sprinkler	____
Painters	____
Tile Setters	____
Carpet Lyrs	____
Controls	____
________	____
________	____
________	____
________	____

SUBCONTRACTORS ON PROJECT & THEIR ACTIVITY

DAILY ACTIVITY - VISITORS - INSPECTIONS

Supervisor's Signature: ______________________________

DAILY LOG

Day: ____________________

November 10, 20____ **Weather: 8:00 A.M. ___ Noon: ___ 4:00 P.M. ___**

WORK FORCE

Supt.	____
Foreman	____
Carpenters	____
Laborers	____
Masons	____
Oper, Engrs.	____
Iron Wrkers	____
Electricians	____
Plumbers	____
Steam Fittrs	____
Sheet Metal	____
Glazers	____
Roofers	____
Sprinkler	____
Painters	____
Tile Setters	____
Carpet Lyrs	____
Controls	____
________	____
________	____
________	____
________	____

SUBCONTRACTORS ON PROJECT & THEIR ACTIVITY

DAILY ACTIVITY - VISITORS - INSPECTIONS

Supervisor's Signature: ______________________________

DAILY LOG

Day: ____________________

November 11, 20____ **Weather: 8:00 A.M. ____ Noon: ____ 4:00 P.M. ____**

WORK FORCE

Supt.	____
Foreman	____
Carpenters	____
Laborers	____
Masons	____
Oper, Engrs.	____
Iron Wrkers	____
Electricians	____
Plumbers	____
Steam Fittrs	____
Sheet Metal	____
Glazers	____
Roofers	____
Sprinkler	____
Painters	____
Tile Setters	____
Carpet Lyrs	____
Controls	____
________	____
________	____
________	____
________	____

SUBCONTRACTORS ON PROJECT & THEIR ACTIVITY

DAILY ACTIVITY - VISITORS - INSPECTIONS

Supervisor's Signature: ______________________________

DAILY LOG

Day: ____________________

November 12, 20____ **Weather: 8:00 A.M. ___ Noon: ___ 4:00 P.M. ___**

WORK FORCE

Supt.	____
Foreman	____
Carpenters	____
Laborers	____
Masons	____
Oper, Engrs.	____
Iron Wrkers	____
Electricians	____
Plumbers	____
Steam Fittrs	____
Sheet Metal	____
Glazers	____
Roofers	____
Sprinkler	____
Painters	____
Tile Setters	____
Carpet Lyrs	____
Controls	____
________	____
________	____
________	____
________	____

SUBCONTRACTORS ON PROJECT & THEIR ACTIVITY

DAILY ACTIVITY - VISITORS - INSPECTIONS

Supervisor's Signature: ______________________________

DAILY LOG

Day: ____________________

November 13, 20____ **Weather: 8:00 A.M. ____ Noon: ____ 4:00 P.M. ____**

WORK FORCE	
Supt.	____
Foreman	____
Carpenters	____
Laborers	____
Masons	____
Oper, Engrs.	____
Iron Wrkers	____
Electricians	____
Plumbers	____
Steam Fittrs	____
Sheet Metal	____
Glazers	____
Roofers	____
Sprinkler	____
Painters	____
Tile Setters	____
Carpet Lyrs	____
Controls	____
________	____
________	____
________	____
________	____

SUBCONTRACTORS ON PROJECT & THEIR ACTIVITY

DAILY ACTIVITY - VISITORS - INSPECTIONS

Supervisor's Signature: ______________________________

DAILY LOG

Day: ____________________

November 14, 20____ **Weather: 8:00 A.M. ___ Noon: ___ 4:00 P.M. ___**

WORK FORCE	
Supt.	___
Foreman	___
Carpenters	___
Laborers	___
Masons	___
Oper, Engrs.	___
Iron Wrkers	___
Electricians	___
Plumbers	___
Steam Fittrs	___
Sheet Metal	___
Glazers	___
Roofers	___
Sprinkler	___
Painters	___
Tile Setters	___
Carpet Lyrs	___
Controls	___
______	___
______	___
______	___
______	___

SUBCONTRACTORS ON PROJECT & THEIR ACTIVITY

DAILY ACTIVITY - VISITORS - INSPECTIONS

Supervisor's Signature: ______________________________

DAILY LOG

Day: ____________________

November 15, 20____ **Weather: 8:00 A.M. ____ Noon: ____ 4:00 P.M. ____**

WORK FORCE

Supt.	____
Foreman	____
Carpenters	____
Laborers	____
Masons	____
Oper, Engrs.	____
Iron Wrkers	____
Electricians	____
Plumbers	____
Steam Fittrs	____
Sheet Metal	____
Glazers	____
Roofers	____
Sprinkler	____
Painters	____
Tile Setters	____
Carpet Lyrs	____
Controls	____
________	____
________	____
________	____
________	____

SUBCONTRACTORS ON PROJECT & THEIR ACTIVITY

DAILY ACTIVITY - VISITORS - INSPECTIONS

Supervisor's Signature: ______________________________

DAILY LOG

Day: ____________________

November 16, 20____ **Weather: 8:00 A.M. ____ Noon: ____ 4:00 P.M. ____**

WORK FORCE	
Supt.	____
Foreman	____
Carpenters	____
Laborers	____
Masons	____
Oper, Engrs.	____
Iron Wrkers	____
Electricians	____
Plumbers	____
Steam Fittrs	____
Sheet Metal	____
Glazers	____
Roofers	____
Sprinkler	____
Painters	____
Tile Setters	____
Carpet Lyrs	____
Controls	____
________	____
________	____
________	____
________	____

SUBCONTRACTORS ON PROJECT & THEIR ACTIVITY

DAILY ACTIVITY - VISITORS - INSPECTIONS

Supervisor's Signature: ______________________________

DAILY LOG

Day: ____________________

November 17, 20____ **Weather: 8:00 A.M. ___ Noon: ___ 4:00 P.M. ___**

WORK FORCE

Supt.	___
Foreman	___
Carpenters	___
Laborers	___
Masons	___
Oper, Engrs.	___
Iron Wrkers	___
Electricians	___
Plumbers	___
Steam Fittrs	___
Sheet Metal	___
Glazers	___
Roofers	___
Sprinkler	___
Painters	___
Tile Setters	___
Carpet Lyrs	___
Controls	___
________	___
________	___
________	___
________	___

SUBCONTRACTORS ON PROJECT & THEIR ACTIVITY

DAILY ACTIVITY - VISITORS - INSPECTIONS

Supervisor's Signature: ______________________________

DAILY LOG

Day: ____________________

November 18, 20____ **Weather: 8:00 A.M. ___ Noon: ___ 4:00 P.M. ___**

WORK FORCE

Supt.	___
Foreman	___
Carpenters	___
Laborers	___
Masons	___
Oper, Engrs.	___
Iron Wrkers	___
Electricians	___
Plumbers	___
Steam Fittrs	___
Sheet Metal	___
Glazers	___
Roofers	___
Sprinkler	___
Painters	___
Tile Setters	___
Carpet Lyrs	___
Controls	___
______	___
______	___
______	___
______	___

SUBCONTRACTORS ON PROJECT & THEIR ACTIVITY

DAILY ACTIVITY - VISITORS - INSPECTIONS

Supervisor's Signature: ______________________________

DAILY LOG

Day: ____________________

November 19, 20____ **Weather: 8:00 A.M. ___ Noon: ___ 4:00 P.M. ___**

WORK FORCE	
Supt.	____
Foreman	____
Carpenters	____
Laborers	____
Masons	____
Oper, Engrs.	____
Iron Wrkers	____
Electricians	____
Plumbers	____
Steam Fittrs	____
Sheet Metal	____
Glazers	____
Roofers	____
Sprinkler	____
Painters	____
Tile Setters	____
Carpet Lyrs	____
Controls	____
________	____
________	____
________	____
________	____

SUBCONTRACTORS ON PROJECT & THEIR ACTIVITY

DAILY ACTIVITY - VISITORS - INSPECTIONS

Supervisor's Signature: ____________________

DAILY LOG

Day: ____________________

November 20, 20____ **Weather: 8:00 A.M. ___ Noon: ___ 4:00 P.M. ___**

WORK FORCE	
Supt.	___
Foreman	___
Carpenters	___
Laborers	___
Masons	___
Oper, Engrs.	___
Iron Wrkers	___
Electricians	___
Plumbers	___
Steam Fittrs	___
Sheet Metal	___
Glazers	___
Roofers	___
Sprinkler	___
Painters	___
Tile Setters	___
Carpet Lyrs	___
Controls	___
______	___
______	___
______	___
______	___

SUBCONTRACTORS ON PROJECT & THEIR ACTIVITY

____

____

____

DAILY ACTIVITY - VISITORS - INSPECTIONS

____

____

____

Supervisor's Signature: ____________________

DAILY LOG

Day: ______________________

November 21, 20____ **Weather: 8:00 A.M. ___ Noon: ___ 4:00 P.M. ___**

WORK FORCE

Supt.	___
Foreman	___
Carpenters	___
Laborers	___
Masons	___
Oper, Engrs.	___
Iron Wrkers	___
Electricians	___
Plumbers	___
Steam Fittrs	___
Sheet Metal	___
Glazers	___
Roofers	___
Sprinkler	___
Painters	___
Tile Setters	___
Carpet Lyrs	___
Controls	___
________	___
________	___
________	___
________	___

SUBCONTRACTORS ON PROJECT & THEIR ACTIVITY

DAILY ACTIVITY - VISITORS - INSPECTIONS

Supervisor's Signature: ______________________

DAILY LOG

Day: ____________________

November 22, 20____ **Weather: 8:00 A.M. ___ Noon: ___ 4:00 P.M. ___**

WORK FORCE	
Supt.	___
Foreman	___
Carpenters	___
Laborers	___
Masons	___
Oper, Engrs.	___
Iron Wrkers	___
Electricians	___
Plumbers	___
Steam Fittrs	___
Sheet Metal	___
Glazers	___
Roofers	___
Sprinkler	___
Painters	___
Tile Setters	___
Carpet Lyrs	___
Controls	___
______	___
______	___
______	___
______	___

SUBCONTRACTORS ON PROJECT & THEIR ACTIVITY

DAILY ACTIVITY - VISITORS - INSPECTIONS

Supervisor's Signature: ____________________

DAILY LOG

Day: ________________

November 23, 20____ **Weather: 8:00 A.M. ___ Noon: ___ 4:00 P.M. ___**

WORK FORCE

Supt.	____
Foreman	____
Carpenters	____
Laborers	____
Masons	____
Oper, Engrs.	____
Iron Wrkers	____
Electricians	____
Plumbers	____
Steam Fittrs	____
Sheet Metal	____
Glazers	____
Roofers	____
Sprinkler	____
Painters	____
Tile Setters	____
Carpet Lyrs	____
Controls	____
________	____
________	____
________	____
________	____

SUBCONTRACTORS ON PROJECT & THEIR ACTIVITY

DAILY ACTIVITY - VISITORS - INSPECTIONS

Supervisor's Signature: ______________________________

DAILY LOG

Day: ____________________

November 24, 20____ **Weather: 8:00 A.M. ___ Noon: ___ 4:00 P.M. ___**

WORK FORCE

Supt.	___
Foreman	___
Carpenters	___
Laborers	___
Masons	___
Oper, Engrs.	___
Iron Wrkers	___
Electricians	___
Plumbers	___
Steam Fittrs	___
Sheet Metal	___
Glazers	___
Roofers	___
Sprinkler	___
Painters	___
Tile Setters	___
Carpet Lyrs	___
Controls	___
________	___
________	___
________	___
________	___

SUBCONTRACTORS ON PROJECT & THEIR ACTIVITY

DAILY ACTIVITY - VISITORS - INSPECTIONS

Supervisor's Signature: ______________________________

DAILY LOG

Day: ________________

November 25, 20____ **Weather: 8:00 A.M. ___ Noon: ___ 4:00 P.M. ___**

WORK FORCE	
Supt.	___
Foreman	___
Carpenters	___
Laborers	___
Masons	___
Oper, Engrs.	___
Iron Wrkers	___
Electricians	___
Plumbers	___
Steam Fittrs	___
Sheet Metal	___
Glazers	___
Roofers	___
Sprinkler	___
Painters	___
Tile Setters	___
Carpet Lyrs	___
Controls	___
______	___
______	___
______	___
______	___

SUBCONTRACTORS ON PROJECT & THEIR ACTIVITY

DAILY ACTIVITY - VISITORS - INSPECTIONS

Supervisor's Signature: ____________________

DAILY LOG

Day: ____________________

November 26, 20____ **Weather: 8:00 A.M. ____ Noon: ____ 4:00 P.M. ____**

WORK FORCE

Supt.	____
Foreman	____
Carpenters	____
Laborers	____
Masons	____
Oper, Engrs.	____
Iron Wrkers	____
Electricians	____
Plumbers	____
Steam Fittrs	____
Sheet Metal	____
Glazers	____
Roofers	____
Sprinkler	____
Painters	____
Tile Setters	____
Carpet Lyrs	____
Controls	____
________	____
________	____
________	____
________	____

SUBCONTRACTORS ON PROJECT & THEIR ACTIVITY

DAILY ACTIVITY - VISITORS - INSPECTIONS

Supervisor's Signature: ____________________

DAILY LOG

Day: ____________________

November 27, 20____ **Weather: 8:00 A.M. ___ Noon: ___ 4:00 P.M. ___**

WORK FORCE	
Supt.	____
Foreman	____
Carpenters	____
Laborers	____
Masons	____
Oper, Engrs.	____
Iron Wrkers	____
Electricians	____
Plumbers	____
Steam Fittrs	____
Sheet Metal	____
Glazers	____
Roofers	____
Sprinkler	____
Painters	____
Tile Setters	____
Carpet Lyrs	____
Controls	____
________	____
________	____
________	____
________	____

SUBCONTRACTORS ON PROJECT & THEIR ACTIVITY

DAILY ACTIVITY - VISITORS - INSPECTIONS

Supervisor's Signature: ______________________________

DAILY LOG

Day: ____________________

November 28, 20____ **Weather: 8:00 A.M. ___ Noon: ___ 4:00 P.M. ___**

WORK FORCE

Supt.	___
Foreman	___
Carpenters	___
Laborers	___
Masons	___
Oper, Engrs.	___
Iron Wrkers	___
Electricians	___
Plumbers	___
Steam Fittrs	___
Sheet Metal	___
Glazers	___
Roofers	___
Sprinkler	___
Painters	___
Tile Setters	___
Carpet Lyrs	___
Controls	___
___	___
___	___
___	___
___	___

SUBCONTRACTORS ON PROJECT & THEIR ACTIVITY

DAILY ACTIVITY - VISITORS - INSPECTIONS

Supervisor's Signature: ______________________________

DAILY LOG

Day: ____________________

November 29, 20____   **Weather: 8:00 A.M. ___ Noon: ___ 4:00 P.M. ___**

WORK FORCE

Supt.	___
Foreman	___
Carpenters	___
Laborers	___
Masons	___
Oper, Engrs.	___
Iron Wrkers	___
Electricians	___
Plumbers	___
Steam Fittrs	___
Sheet Metal	___
Glazers	___
Roofers	___
Sprinkler	___
Painters	___
Tile Setters	___
Carpet Lyrs	___
Controls	___
______	___
______	___
______	___
______	___

SUBCONTRACTORS ON PROJECT & THEIR ACTIVITY

DAILY ACTIVITY - VISITORS - INSPECTIONS

Supervisor's Signature: ______________________________

DAILY LOG

Day: ____________________

November 30, 20____ **Weather: 8:00 A.M.** ___ **Noon:** ___ **4:00 P.M.** ___

WORK FORCE

Supt.	___
Foreman	___
Carpenters	___
Laborers	___
Masons	___
Oper, Engrs.	___
Iron Wrkers	___
Electricians	___
Plumbers	___
Steam Fittrs	___
Sheet Metal	___
Glazers	___
Roofers	___
Sprinkler	___
Painters	___
Tile Setters	___
Carpet Lyrs	___
Controls	___
___	___
___	___
___	___
___	___

SUBCONTRACTORS ON PROJECT & THEIR ACTIVITY

DAILY ACTIVITY - VISITORS - INSPECTIONS

Supervisor's Signature: ____________________

DAILY LOG

Day: ____________________

December 1, 20____ **Weather: 8:00 A.M. ____ Noon: ____ 4:00 P.M. ____**

WORK FORCE

Supt.	____
Foreman	____
Carpenters	____
Laborers	____
Masons	____
Oper, Engrs.	____
Iron Wrkers	____
Electricians	____
Plumbers	____
Steam Fittrs	____
Sheet Metal	____
Glazers	____
Roofers	____
Sprinkler	____
Painters	____
Tile Setters	____
Carpet Lyrs	____
Controls	____
________	____
________	____
________	____
________	____

SUBCONTRACTORS ON PROJECT & THEIR ACTIVITY

DAILY ACTIVITY - VISITORS - INSPECTIONS

Supervisor's Signature: ______________________________

DAILY LOG

Day: ____________________

December 2, 20____ **Weather: 8:00 A.M. ____ Noon: ____ 4:00 P.M. ____**

WORK FORCE

Supt.	____
Foreman	____
Carpenters	____
Laborers	____
Masons	____
Oper, Engrs.	____
Iron Wrkers	____
Electricians	____
Plumbers	____
Steam Fittrs	____
Sheet Metal	____
Glazers	____
Roofers	____
Sprinkler	____
Painters	____
Tile Setters	____
Carpet Lyrs	____
Controls	____
________	____
________	____
________	____
________	____

SUBCONTRACTORS ON PROJECT & THEIR ACTIVITY

DAILY ACTIVITY - VISITORS - INSPECTIONS

Supervisor's Signature: ______________________________

DAILY LOG

Day: ____________________

December 3, 20____ **Weather: 8:00 A.M. ___ Noon: ___ 4:00 P.M. ___**

WORK FORCE

Supt.	____
Foreman	____
Carpenters	____
Laborers	____
Masons	____
Oper, Engrs.	____
Iron Wrkers	____
Electricians	____
Plumbers	____
Steam Fittrs	____
Sheet Metal	____
Glazers	____
Roofers	____
Sprinkler	____
Painters	____
Tile Setters	____
Carpet Lyrs	____
Controls	____
________	____
________	____
________	____
________	____

SUBCONTRACTORS ON PROJECT & THEIR ACTIVITY

DAILY ACTIVITY - VISITORS - INSPECTIONS

Supervisor's Signature: ______________________________

DAILY LOG

Day: ____________________

December 4, 20____ **Weather: 8:00 A.M. ___ Noon: ___ 4:00 P.M. ___**

WORK FORCE

Supt.	___
Foreman	___
Carpenters	___
Laborers	___
Masons	___
Oper, Engrs.	___
Iron Wrkers	___
Electricians	___
Plumbers	___
Steam Fittrs	___
Sheet Metal	___
Glazers	___
Roofers	___
Sprinkler	___
Painters	___
Tile Setters	___
Carpet Lyrs	___
Controls	___
______	___
______	___
______	___
______	___

SUBCONTRACTORS ON PROJECT & THEIR ACTIVITY

DAILY ACTIVITY - VISITORS - INSPECTIONS

Supervisor's Signature: ______________________________

DAILY LOG

Day: ____________________

December 5, 20____ **Weather: 8:00 A.M. ___ Noon: ___ 4:00 P.M. ___**

WORK FORCE

Supt.	____
Foreman	____
Carpenters	____
Laborers	____
Masons	____
Oper, Engrs.	____
Iron Wrkers	____
Electricians	____
Plumbers	____
Steam Fittrs	____
Sheet Metal	____
Glazers	____
Roofers	____
Sprinkler	____
Painters	____
Tile Setters	____
Carpet Lyrs	____
Controls	____
________	____
________	____
________	____
________	____

SUBCONTRACTORS ON PROJECT & THEIR ACTIVITY

DAILY ACTIVITY - VISITORS - INSPECTIONS

Supervisor's Signature: ______________________________

DAILY LOG

Day: ____________________

December 6, 20____ **Weather: 8:00 A.M. ___ Noon: ___ 4:00 P.M. ___**

WORK FORCE

Supt.	___
Foreman	___
Carpenters	___
Laborers	___
Masons	___
Oper, Engrs.	___
Iron Wrkers	___
Electricians	___
Plumbers	___
Steam Fittrs	___
Sheet Metal	___
Glazers	___
Roofers	___
Sprinkler	___
Painters	___
Tile Setters	___
Carpet Lyrs	___
Controls	___
___	___
___	___
___	___
___	___

SUBCONTRACTORS ON PROJECT & THEIR ACTIVITY

DAILY ACTIVITY - VISITORS - INSPECTIONS

Supervisor's Signature: ______________________________

DAILY LOG

Day: ____________________

December 7, 20____ **Weather: 8:00 A.M. ___ Noon: ___ 4:00 P.M. ___**

WORK FORCE

Supt.	___
Foreman	___
Carpenters	___
Laborers	___
Masons	___
Oper, Engrs.	___
Iron Wrkers	___
Electricians	___
Plumbers	___
Steam Fittrs	___
Sheet Metal	___
Glazers	___
Roofers	___
Sprinkler	___
Painters	___
Tile Setters	___
Carpet Lyrs	___
Controls	___
________	___
________	___
________	___
________	___

SUBCONTRACTORS ON PROJECT & THEIR ACTIVITY

DAILY ACTIVITY - VISITORS - INSPECTIONS

Supervisor's Signature: ____________________

DAILY LOG

Day: ____________________

December 8, 20____ **Weather: 8:00 A.M. ___ Noon: ___ 4:00 P.M. ___**

WORK FORCE

Supt.	___
Foreman	___
Carpenters	___
Laborers	___
Masons	___
Oper, Engrs.	___
Iron Wrkers	___
Electricians	___
Plumbers	___
Steam Fittrs	___
Sheet Metal	___
Glazers	___
Roofers	___
Sprinkler	___
Painters	___
Tile Setters	___
Carpet Lyrs	___
Controls	___
________	___
________	___
________	___
________	___

SUBCONTRACTORS ON PROJECT & THEIR ACTIVITY

DAILY ACTIVITY - VISITORS - INSPECTIONS

Supervisor's Signature: ______________________________

DAILY LOG

Day: ____________________

December 9, 20____ **Weather: 8:00 A.M. ____ Noon: ____ 4:00 P.M. ____**

WORK FORCE

Supt.	____
Foreman	____
Carpenters	____
Laborers	____
Masons	____
Oper, Engrs.	____
Iron Wrkers	____
Electricians	____
Plumbers	____
Steam Fittrs	____
Sheet Metal	____
Glazers	____
Roofers	____
Sprinkler	____
Painters	____
Tile Setters	____
Carpet Lyrs	____
Controls	____
________	____
________	____
________	____
________	____

SUBCONTRACTORS ON PROJECT & THEIR ACTIVITY

DAILY ACTIVITY - VISITORS - INSPECTIONS

Supervisor's Signature: ______________________________

DAILY LOG

Day: ____________________

December 10, 20____ **Weather: 8:00 A.M.** ____ **Noon:** ____ **4:00 P.M.** ____

WORK FORCE

Supt.	____
Foreman	____
Carpenters	____
Laborers	____
Masons	____
Oper, Engrs.	____
Iron Wrkers	____
Electricians	____
Plumbers	____
Steam Fittrs	____
Sheet Metal	____
Glazers	____
Roofers	____
Sprinkler	____
Painters	____
Tile Setters	____
Carpet Lyrs	____
Controls	____
____	____
____	____
____	____
____	____

SUBCONTRACTORS ON PROJECT & THEIR ACTIVITY

DAILY ACTIVITY - VISITORS - INSPECTIONS

Supervisor's Signature: ______________________________

DAILY LOG

Day: ____________________

December 11, 20____ **Weather: 8:00 A.M. ____ Noon: ____ 4:00 P.M. ____**

WORK FORCE	
Supt.	___
Foreman	___
Carpenters	___
Laborers	___
Masons	___
Oper, Engrs.	___
Iron Wrkers	___
Electricians	___
Plumbers	___
Steam Fittrs	___
Sheet Metal	___
Glazers	___
Roofers	___
Sprinkler	___
Painters	___
Tile Setters	___
Carpet Lyrs	___
Controls	___
___	___
___	___
___	___
___	___

SUBCONTRACTORS ON PROJECT & THEIR ACTIVITY

DAILY ACTIVITY - VISITORS - INSPECTIONS

Supervisor's Signature: ____________________

DAILY LOG

Day: ____________________

December 12, 20____ **Weather: 8:00 A.M. ___ Noon: ___ 4:00 P.M. ___**

WORK FORCE

Supt.	____
Foreman	____
Carpenters	____
Laborers	____
Masons	____
Oper, Engrs.	____
Iron Wrkers	____
Electricians	____
Plumbers	____
Steam Fittrs	____
Sheet Metal	____
Glazers	____
Roofers	____
Sprinkler	____
Painters	____
Tile Setters	____
Carpet Lyrs	____
Controls	____
________	____
________	____
________	____
________	____

SUBCONTRACTORS ON PROJECT & THEIR ACTIVITY

DAILY ACTIVITY - VISITORS - INSPECTIONS

Supervisor's Signature: ______________________________

DAILY LOG

Day: ____________________

December 13, 20____ **Weather: 8:00 A.M. ___ Noon: ___ 4:00 P.M. ___**

WORK FORCE	
Supt.	___
Foreman	___
Carpenters	___
Laborers	___
Masons	___
Oper, Engrs.	___
Iron Wrkers	___
Electricians	___
Plumbers	___
Steam Fittrs	___
Sheet Metal	___
Glazers	___
Roofers	___
Sprinkler	___
Painters	___
Tile Setters	___
Carpet Lyrs	___
Controls	___
________	___
________	___
________	___
________	___

SUBCONTRACTORS ON PROJECT & THEIR ACTIVITY

DAILY ACTIVITY - VISITORS - INSPECTIONS

Supervisor's Signature: ______________________________

DAILY LOG

Day: ____________________

December 14, 20____   **Weather: 8:00 A.M. ____ Noon: ____ 4:00 P.M. ____**

WORK FORCE

Supt.	____
Foreman	____
Carpenters	____
Laborers	____
Masons	____
Oper, Engrs.	____
Iron Wrkers	____
Electricians	____
Plumbers	____
Steam Fittrs	____
Sheet Metal	____
Glazers	____
Roofers	____
Sprinkler	____
Painters	____
Tile Setters	____
Carpet Lyrs	____
Controls	____
________	____
________	____
________	____
________	____

SUBCONTRACTORS ON PROJECT & THEIR ACTIVITY

DAILY ACTIVITY - VISITORS - INSPECTIONS

Supervisor's Signature: ______________________________

DAILY LOG

Day: ____________________

December 15, 20____ **Weather: 8:00 A.M. ___ Noon: ___ 4:00 P.M. ___**

WORK FORCE	
Supt.	___
Foreman	___
Carpenters	___
Laborers	___
Masons	___
Oper, Engrs.	___
Iron Wrkers	___
Electricians	___
Plumbers	___
Steam Fittrs	___
Sheet Metal	___
Glazers	___
Roofers	___
Sprinkler	___
Painters	___
Tile Setters	___
Carpet Lyrs	___
Controls	___
________	___
________	___
________	___
________	___

SUBCONTRACTORS ON PROJECT & THEIR ACTIVITY

DAILY ACTIVITY - VISITORS - INSPECTIONS

Supervisor's Signature: ______________________________

DAILY LOG

Day: ____________________

December 16, 20____ **Weather: 8:00 A.M. ___ Noon: ___ 4:00 P.M. ___**

WORK FORCE

Supt.	___
Foreman	___
Carpenters	___
Laborers	___
Masons	___
Oper, Engrs.	___
Iron Wrkers	___
Electricians	___
Plumbers	___
Steam Fittrs	___
Sheet Metal	___
Glazers	___
Roofers	___
Sprinkler	___
Painters	___
Tile Setters	___
Carpet Lyrs	___
Controls	___
______	___
______	___
______	___
______	___

SUBCONTRACTORS ON PROJECT & THEIR ACTIVITY

DAILY ACTIVITY - VISITORS - INSPECTIONS

Supervisor's Signature: ______________________________

DAILY LOG

Day: ________________

December 17, 20____ **Weather: 8:00 A.M.** ___ **Noon:** ___ **4:00 P.M.** ___

WORK FORCE	
Supt.	___
Foreman	___
Carpenters	___
Laborers	___
Masons	___
Oper, Engrs.	___
Iron Wrkers	___
Electricians	___
Plumbers	___
Steam Fittrs	___
Sheet Metal	___
Glazers	___
Roofers	___
Sprinkler	___
Painters	___
Tile Setters	___
Carpet Lyrs	___
Controls	___
______	___
______	___
______	___
______	___

SUBCONTRACTORS ON PROJECT & THEIR ACTIVITY

DAILY ACTIVITY - VISITORS - INSPECTIONS

Supervisor's Signature: ______________________________

DAILY LOG

Day: ____________________

December 18, 20____ **Weather: 8:00 A.M. ___ Noon: ___ 4:00 P.M. ___**

WORK FORCE

Supt.	___
Foreman	___
Carpenters	___
Laborers	___
Masons	___
Oper, Engrs.	___
Iron Wrkers	___
Electricians	___
Plumbers	___
Steam Fittrs	___
Sheet Metal	___
Glazers	___
Roofers	___
Sprinkler	___
Painters	___
Tile Setters	___
Carpet Lyrs	___
Controls	___
______	___
______	___
______	___
______	___

SUBCONTRACTORS ON PROJECT & THEIR ACTIVITY

DAILY ACTIVITY - VISITORS - INSPECTIONS

Supervisor's Signature: ______________________________

DAILY LOG

Day: ____________________

December 19, 20____ **Weather: 8:00 A.M. ____ Noon: ____ 4:00 P.M. ____**

WORK FORCE

Supt.	____
Foreman	____
Carpenters	____
Laborers	____
Masons	____
Oper, Engrs.	____
Iron Wrkers	____
Electricians	____
Plumbers	____
Steam Fittrs	____
Sheet Metal	____
Glazers	____
Roofers	____
Sprinkler	____
Painters	____
Tile Setters	____
Carpet Lyrs	____
Controls	____
________	____
________	____
________	____
________	____

SUBCONTRACTORS ON PROJECT & THEIR ACTIVITY

DAILY ACTIVITY - VISITORS - INSPECTIONS

Supervisor's Signature: ______________________________

DAILY LOG

Day: ____________________

December 20, 20____ **Weather: 8:00 A.M. ___ Noon: ___ 4:00 P.M. ___**

WORK FORCE	
Supt.	____
Foreman	____
Carpenters	____
Laborers	____
Masons	____
Oper, Engrs.	____
Iron Wrkers	____
Electricians	____
Plumbers	____
Steam Fittrs	____
Sheet Metal	____
Glazers	____
Roofers	____
Sprinkler	____
Painters	____
Tile Setters	____
Carpet Lyrs	____
Controls	____
________	____
________	____
________	____
________	____

SUBCONTRACTORS ON PROJECT & THEIR ACTIVITY

DAILY ACTIVITY - VISITORS - INSPECTIONS

Supervisor's Signature: ____________________

DAILY LOG

Day: ________________

December 21, 20____ **Weather: 8:00 A.M. ___ Noon: ___ 4:00 P.M. ___**

WORK FORCE

Supt.	___
Foreman	___
Carpenters	___
Laborers	___
Masons	___
Oper, Engrs.	___
Iron Wrkers	___
Electricians	___
Plumbers	___
Steam Fittrs	___
Sheet Metal	___
Glazers	___
Roofers	___
Sprinkler	___
Painters	___
Tile Setters	___
Carpet Lyrs	___
Controls	___
______	___
______	___
______	___
______	___

SUBCONTRACTORS ON PROJECT & THEIR ACTIVITY

DAILY ACTIVITY - VISITORS - INSPECTIONS

Supervisor's Signature: ______________________________

DAILY LOG

Day: ____________________

December 22, 20____ **Weather: 8:00 A.M. ___ Noon: ___ 4:00 P.M. ___**

WORK FORCE		**SUBCONTRACTORS ON PROJECT & THEIR ACTIVITY**
Supt.	____	
Foreman	____	
Carpenters	____	
Laborers	____	
Masons	____	
Oper, Engrs.	____	
Iron Wrkers	____	
Electricians	____	
Plumbers	____	
Steam Fittrs	____	
Sheet Metal	____	
Glazers	____	
Roofers	____	
Sprinkler	____	
Painters	____	
Tile Setters	____	
Carpet Lyrs	____	
Controls	____	
________	____	DAILY ACTIVITY - VISITORS - INSPECTIONS
________	____	
________	____	
________	____	

Supervisor's Signature: ________________________________

DAILY LOG

Day: ____________________

December 23, 20____ **Weather: 8:00 A.M. ___ Noon: ___ 4:00 P.M. ___**

WORK FORCE

Supt.	___
Foreman	___
Carpenters	___
Laborers	___
Masons	___
Oper, Engrs.	___
Iron Wrkers	___
Electricians	___
Plumbers	___
Steam Fittrs	___
Sheet Metal	___
Glazers	___
Roofers	___
Sprinkler	___
Painters	___
Tile Setters	___
Carpet Lyrs	___
Controls	___
________	___
________	___
________	___
________	___

SUBCONTRACTORS ON PROJECT & THEIR ACTIVITY

DAILY ACTIVITY - VISITORS - INSPECTIONS

Supervisor's Signature: ________________________________

DAILY LOG

Day: ____________________

December 24, 20____ **Weather: 8:00 A.M. ____ Noon: ____ 4:00 P.M. ____**

WORK FORCE

Supt.	____
Foreman	____
Carpenters	____
Laborers	____
Masons	____
Oper, Engrs.	____
Iron Wrkers	____
Electricians	____
Plumbers	____
Steam Fittrs	____
Sheet Metal	____
Glazers	____
Roofers	____
Sprinkler	____
Painters	____
Tile Setters	____
Carpet Lyrs	____
Controls	____
________	____
________	____
________	____
________	____

SUBCONTRACTORS ON PROJECT & THEIR ACTIVITY

DAILY ACTIVITY - VISITORS - INSPECTIONS

Supervisor's Signature: ______________________________

DAILY LOG

Day: ____________________

December 25, 20____ **Weather: 8:00 A.M. ____ Noon: ____ 4:00 P.M. ____**

WORK FORCE	
Supt.	____
Foreman	____
Carpenters	____
Laborers	____
Masons	____
Oper, Engrs.	____
Iron Wrkers	____
Electricians	____
Plumbers	____
Steam Fittrs	____
Sheet Metal	____
Glazers	____
Roofers	____
Sprinkler	____
Painters	____
Tile Setters	____
Carpet Lyrs	____
Controls	____
________	____
________	____
________	____
________	____

SUBCONTRACTORS ON PROJECT & THEIR ACTIVITY

DAILY ACTIVITY - VISITORS - INSPECTIONS

Supervisor's Signature: ________________________________

DAILY LOG

Day: ____________

December 26, 20____ **Weather: 8:00 A.M. ___ Noon: ___ 4:00 P.M. ___**

WORK FORCE

Supt.	___
Foreman	___
Carpenters	___
Laborers	___
Masons	___
Oper, Engrs.	___
Iron Wrkers	___
Electricians	___
Plumbers	___
Steam Fittrs	___
Sheet Metal	___
Glazers	___
Roofers	___
Sprinkler	___
Painters	___
Tile Setters	___
Carpet Lyrs	___
Controls	___
______	___
______	___
______	___
______	___

SUBCONTRACTORS ON PROJECT & THEIR ACTIVITY

DAILY ACTIVITY - VISITORS - INSPECTIONS

Supervisor's Signature: ____________________

DAILY LOG

Day: ____________________

December 27, 20____ **Weather: 8:00 A.M. ___ Noon: ___ 4:00 P.M. ___**

WORK FORCE		SUBCONTRACTORS ON PROJECT & THEIR ACTIVITY
Supt.	____	
Foreman	____	
Carpenters	____	
Laborers	____	
Masons	____	
Oper, Engrs.	____	
Iron Wrkers	____	
Electricians	____	
Plumbers	____	
Steam Fittrs	____	
Sheet Metal	____	
Glazers	____	
Roofers	____	
Sprinkler	____	
Painters	____	
Tile Setters	____	
Carpet Lyrs	____	
Controls	____	
________	____	DAILY ACTIVITY - VISITORS - INSPECTIONS
________	____	
________	____	
________	____	

Supervisor's Signature: ______________________________

DAILY LOG

Day: ____________________

December 28, 20____ **Weather: 8:00 A.M. ____ Noon: ____ 4:00 P.M. ____**

WORK FORCE

Supt.	____
Foreman	____
Carpenters	____
Laborers	____
Masons	____
Oper, Engrs.	____
Iron Wrkers	____
Electricians	____
Plumbers	____
Steam Fittrs	____
Sheet Metal	____
Glazers	____
Roofers	____
Sprinkler	____
Painters	____
Tile Setters	____
Carpet Lyrs	____
Controls	____
________	____
________	____
________	____
________	____

SUBCONTRACTORS ON PROJECT & THEIR ACTIVITY

DAILY ACTIVITY - VISITORS - INSPECTIONS

Supervisor's Signature: ______________________________

DAILY LOG

Day: ____________________

December 29, 20____ **Weather: 8:00 A.M. ___ Noon: ___ 4:00 P.M. ___**

WORK FORCE	
Supt.	___
Foreman	___
Carpenters	___
Laborers	___
Masons	___
Oper, Engrs.	___
Iron Wrkers	___
Electricians	___
Plumbers	___
Steam Fittrs	___
Sheet Metal	___
Glazers	___
Roofers	___
Sprinkler	___
Painters	___
Tile Setters	___
Carpet Lyrs	___
Controls	___
___	___
___	___
___	___
___	___

SUBCONTRACTORS ON PROJECT & THEIR ACTIVITY

DAILY ACTIVITY - VISITORS - INSPECTIONS

Supervisor's Signature: ____________________

DAILY LOG

Day: ____________________

December 30, 20____ **Weather: 8:00 A.M. ___ Noon: ___ 4:00 P.M. ___**

WORK FORCE

Supt.	____
Foreman	____
Carpenters	____
Laborers	____
Masons	____
Oper, Engrs.	____
Iron Wrkers	____
Electricians	____
Plumbers	____
Steam Fittrs	____
Sheet Metal	____
Glazers	____
Roofers	____
Sprinkler	____
Painters	____
Tile Setters	____
Carpet Lyrs	____
Controls	____
________	____
________	____
________	____
________	____

SUBCONTRACTORS ON PROJECT & THEIR ACTIVITY

DAILY ACTIVITY - VISITORS - INSPECTIONS

Supervisor's Signature: ____________________

DAILY LOG

Day: ____________________

December 31, 20____ **Weather: 8:00 A.M. ____ Noon: ____ 4:00 P.M. ____**

WORK FORCE

Supt.	____
Foreman	____
Carpenters	____
Laborers	____
Masons	____
Oper, Engrs.	____
Iron Wrkers	____
Electricians	____
Plumbers	____
Steam Fittrs	____
Sheet Metal	____
Glazers	____
Roofers	____
Sprinkler	____
Painters	____
Tile Setters	____
Carpet Lyrs	____
Controls	____
________	____
________	____
________	____
________	____

SUBCONTRACTORS ON PROJECT & THEIR ACTIVITY

DAILY ACTIVITY - VISITORS - INSPECTIONS

Supervisor's Signature: ______________________________

www.ingramcontent.com/pod-product-compliance
Lightning Source LLC
LaVergne TN
LVHW080010210926
848481LV00011B/2982
9780071408141